Learning and Improvement Curves

Arkie Fanning PE, CCE, LQA

Learning and Improvement Curves

Chapter 1: Basics

Basic Theory: The basic theory of learning curves is simple. The more people do a particular job the better they become at doing that particular job and therefore it will take less time to do that job with each repetition. When learning curves are used to describe processes they are usually called improvement curves. Basically, the more often a process is performed, the more likely people are to come up with a better way of doing that process. The math used to calculate improvement curves is the same math used to calculate learning curves and in this text, every method used to calculate a particular facet of a learning curve can also be used to calculate a similar facet in an improvement curve.

Types of learning curves: There are two types of learning curves commonly in use. Both types use the same mathematical model, however there is a different meaning attached to one variable in that model.

The most commonly used learning curve is called the unit learning curve (or the Crawford curve). The unit curve estimates the cost/time for the particular unit under investigation. For example, this model would provide the following answer, the time required to process the 8th unit is 3.25 hours. The basic equation used in the unit learning curve model is:

$Y = A(x^b)$ where:

Y = Unit time for the xth unit
A = Time required to produce the 1st unit (also called T1)
b = The learning exponent

Each of these variables will be discussed in full later but first let's look at the other commonly used learning curve.

The other commonly used learning curve is called the cumulative average (or the Wright curve). The cumulative average curve estimates the cost/time for the cumulative average of all units under investigation. For example, this model would provide the following answer, the average time required to process the first ten units is 2.45 hours per unit. The basic equation used in the cumulative average learning curve is:

$Y = A(x^b)$ where:

Y = Cumulative average time for the xth unit
A = Time required to produce the 1st unit (also called T1)
b = The learning exponent

To illustrate the difference in these two methods let us take the same set of assumptions and determine what the answers would be under each methodology.

Let $x = 2$
Let $A = 100$ hours
Let $b = -0.152$ (this represents a 90% learning curve and will be fully discussed later)

For the unit learning curve the 2nd unit time is :

$Y = A(x^b) = 100(2^{-.152}) = 90$ hours

For the cumulative average learning curve the cumulative average time for the first two units is:

$Y = A(x^b) = 100(2^{-.152}) = 90$ hours

So where is the difference?

The total estimated time required to process the first two units according to the unit learning curve model is the time of the first unit (100 hours) plus the time of the second unit (90 hours) for a total time of 190 hours or an average time of 190/2 = 95 hours per unit

 The total estimated time required to process the first two units according to the cumulative average learning curve model is the time of the second unit (90 hours) multiplied by the total number of units (2) for a total time of 180 hours/2 = 90 hours per unit

For some reason the unit curve model (although more cumbersome in my opinion) has become the most used model and it will be the model most discussed in this text. It is important to remember that the methods are simply conventions and they will provide equivalent answers for a given data set. For example:

The first unit time is 100 hours, the second unit time is 90 hours what are the unit and cum average learning curves that would be associated with this data set?

Unit curve: This represents a 90% learning curve

The predicted cost would have been 100 hrs for the first unit, and 90 hours for the second unit for a total of 190 hours and a cumulative average cost of 95 hours per unit.

e.g.

$Y_1 = A(x^b) = 100(1^{-.152}) = 100$ hours
$Y_2 = A(x^b) = 100(2^{-.152}) = 90$ hours

Total $Y_{1+2} = 190$ hours

Cumulative Average Curve: This represents a 95% cum average learning curve

$Y_{1+2} = A(x^b) = 100(2^{-.07399}) = 95$ hours

Total $Y_{1+2} = 95 \times 2 = 190$ hours

As you can see, the same data can be used to determine the appropriate cum average or unit learning curve. However, the estimates for the future will be different, let's continue the example with

our 90% unit and 95% cum average
learning curves:

Unit learning curve prediction of the third
unit:

$Y_3 = A(x^b) = 100(3^{-.152}) = 84.62$ hours

Total $Y_{1+2+3} = 100 + 90 + 84.62 = 274.62$
hours and the average time per unit is =
274.62/3 = 91.54 hours/unit

The cumulative average learning curve
prediction of the first three units

$Y_{1+2+3} = A(x^b) = 100(3^{-.07399}) = 92.19$
hours/unit

There is a difference in prediction of 0.65
hours per unit between the two methods
in the projection. Which is correct? Take
your choice. If we tracked the data and
found that we actually required 80 hours
to process the third unit then we would
find that our learning curves based on the
real data would change for both methods.

 The point of the above is that the
selection of which method to use has no
impact on the actual learning that goes
on. Either method will capture the
learning and couch the data in the terms
of that model. Use the one you feel

comfortable with (or the one your boss or contract says to use).

Most of this text will be directed to unit learning curves because they are (by far) the most used in the defence industry (where I have spent the majority of my career). Just remember that for every mathematical method that has been developed to assist us in understanding the unit model, there is a similar method that can be used to assist us in understanding the cumulative average model.

Learning curve variables: There are four variables used in the determination of learning curves. If you have only one data set you need three of the variables to predict the fourth. The variables are:

$Y = A(x^b)$ where:

$Y =$ Time required to produce xth unit
$A =$ Time required to produce the 1st unit (also called T1)
$x =$ The unit being produced
$b =$ The learning exponent

Y: Y is the time required to produce the unit being investigated. It can also be the cost to produce the unit being investigated. Y is often known for the past and unknown for the future and we are trying to use our best methods to determine what the future Y will be.

A: A is the time it takes to produce the first unit. It can also be the cost required to produce the first unit. Please note that cost is a perfectly good substitute for time and is used extensively in learning curve or improvement curve situations. A is usually **unknown**. Let me repeat, A is usually **unknown**. In most real situations, cost and or time tracking is not to the requisite degree to determine A and usually A has to be determined after the fact.

A is also called T1. T1 stands for touch labor for the first unit.

X: x is the unit under investigation. When lot calculations are required in unit learning theory (an area that will be discussed ad nauseum later in this text), x is represented by the midpoint of the lot and will be referred to as x_{mid}. X is almost always known.

b: b is the learning exponent and it is a mathematical estimate of the amount of learning that will take place. It is calculated by taking the log of the learning curve and dividing it by log 2. b is seldom known. Usually either an assumption is made about b or else the purpose of the analysis is to determine b. Here's an example of how to calculate b:

I want to know what the unit learning exponent is for a 90% learning curve.

First take the $\log_{10}$ of .90 = -0.04576
Then take the $\log_{10}$ of 2 = 0.3010

$b=\log_{10}.90/\log_{10}2$= -0.04576/0.3010 = - 0.152

What this says is that the time to do the second unit is 90% of the time it took to do the first unit. The time to do the fourth (not the third) unit is 90% of the time it took to do the second unit. Take this out as far as you want to go.

For example:

Let A = 100 hours

Then Y_1 = 100 hours
Y_2= .90 X 100 = 90 hours
Y_4= .90 X 90 = 81 hours

Or we could calculate the 4th unit directly

$Y = A(x^b)= 100(4^{-.152})= 81$

For instance what would be the time estimated for the 8th unit?

$$Y_8 = .90 \times Y_4 = .90 \times 81 = 72.9 \text{ or}$$

$$Y = A(x^b) = 100(8^{-.152}) = 72.9$$

As you can see, every time our rate doubles it takes us 10% less time to do the unit in question. Why?

Because the learning exponent, b, is made up of two parts. The learning rate, in this case 90% and the doubling rate which is always 2. There is nothing magical about this, it was just a best fit to the data that was used in the original learning curve papers. Another form might work just as well and if you are looking for a thesis or dissertation you might choose to delve into this.

You will see b written as –b in some texts. The leading minus sign is not necessary because when you do the appropriate calculations the b value will become negative.

Examples of simple learning curve calculations:

Following are examples that show how to calculate learning curves given that you have 3 variables known and one unknown.

I have the following information:

Solving for y

Joe the engineer told me that it took 40 hours to make the first unit of our new widgets. Jan the manufacturing finance person told me that we had already built 8 widgets and we are soon to start on the ninth. I estimate from experience that we can obtain an 85% learning curve on our work. My boss comes by and asked me to tell him how many hours will be required to produce units 9 through 12.

First, what am I going to estimate?

I am going to estimate the time it takes to produce units Y_9, Y_{10}, Y_{11}, and Y_{12}.

What information do I have?

First, A or T1 = 40 hours (You should be so lucky as to have this information)

X = units 9, 10, 11 and 12

$b = \log_{10}(.85)/\log_{10}(2) = -0.0706/0.301 = -0.2345$

$Y_{9+10+11+12} = A(x_9{}^b) + A(x_{10}{}^b) + A(x_{11}{}^b) + A(x_{12}{}^b) =$

$40(9^{-.2345}) + 40(10^{-.2345}) + 40(11^{-.2345}) + 40(12^{-.2345}) =$

$23.90 + 23.31 + 22.80 + 22.34 = 92.34$

Units 9, 10, 11 and 12 will take an estimated 92.34 hours to process.

Solving for A

My company is entering into negotiations to develop a new product. This product is similar to another product that we currently sell. The boss asks me to develop a labor estimate for the labor required to build the first product on the new product line. I know the following information from the old product:

We have produced 200 of these widgets. The recorded time required to produce the 200th unit was 38.2 hours. There are no records available that tell me the first unit labor. We anticipate the appropriate learning curve for the new product will be 90%.

What do I know:

$Y_{200} = 38.2$

$X = 200$

$b = \log_{10}(.90)/\log_{10}(2) = -0.045/0.301 = -0.152$

Set up the first equation:

$Y_{200} = A(200^b) = A(200^{-0.152}) = .447A$

We know that Y_{200} for the old product is 38.2

The new equation becomes:

$38.2 = .447A$

$A = 85.5$ hours

By analogy I estimate the time required for the first unit of the new product will be 85.5 hours.

Solving for X

Rarely, if ever, you might be asked to determine when production will hit a certain rate. The question might be: When will I be able to hit a 40 hour time required to produce a unit?

Let's say the following information is available:

A = 200 hours
Y = 40 (as given above)
X = unknown
LC = 90%

First solve for b: $b = \log_{10}(.90)/\log_{10}(2) = -0.045/0.301 = -0.152$

Set up the equation

$$Y = A(x^b)$$

$$40 = 200(x^{-.152})$$

Now comes the tricky part. You can solve this on a spreadsheet by simply plugging in numbers until the x value is revealed, but to solve by algebra you must first make the equation linear. To do this, take logarithms for both sides:

Recall that $Y = A(x^b)$ expressed in logarithms is:

$$\text{Log } Y = \text{Log } A + b \,(\text{Log } x)$$

Remember this equation, you will see it or something resembling it many times in this text.

Now take the logs:

Log Y = $Log_{10}40$ = 1.60

Log A = $Log_{10}200$ = 2.30

b = -0.152

Let's put this into our equation:

1.60 = 2.30 -.152(Log x)

-0.70 = -0.152(Log x)

Log x = -.70/-0.152 = 4.60

Anti Log x = $10^{4.605}$ = 39,663

So unit 39,663 or there about you will hit a 40 hour per unit time

You may get slightly different values with your hand held calculator due to rounding but the answer should be close to 40,000.

In all my years of using learning curves (many more than I would care to admit) I don't think I have had to solve for x more than a few times. However, it is easily done once you have the proper data.

A common question that has been asked of me over the years is why I use logarithms with base 10 rather than natural logarithms. The answer is: there is no reason. You can do the logarithms in any base and as long as you stay consistent with their use you will get the same answers.

Solving for b:

You have the following information:

Y = 90 hours
A = 200 hours
X = unit 30

What learning curve does this represent?

Set up the equation:

$Y = A(x^b)$

$90 = 200(30^b)$

Again you can solve via a spreadsheet or you can use the same trick I showed you above.

$Log_{10}90 = Log_{10}200 + b(Log_{10}30)$

$1.95 = 2.30 + b(1.48)$

-0.35 = 1.48b

b = -0.35/1.48 = -0.2365

recall that b is composed of the learning rate and the doubling rate and the doubling rate is always the logarithm of 2

b = learning rate/$\log_{10}2$ = learning rate/0.301

-0.2365 = learning rate/0.301

learning rate = -0.2365 * 0.301 = -0.071

The learning curve is the antilog of the learning rate therefore

learning curve = $10^{-0.071}$ = 0.85

The learning curve for this data set is 0.85 or when expressed as a percentage (which it almost always is) 85%.

Solving for learning curve variables when unit information is known.

Suppose you have the following information:

Unit 12: $326

Unit 24: $283

Can you determine the learning curve variables (i.e. A, LC % and b)?

Easily. Here's how.

$Y_{12} = A(12^b)$

$Y_{24} = A(24^b)$

$Y_{12}/Y_{24} = A(12^b)/ A(24^b)$

$326/283 = (12^b)/(24^b)$ (Eliminate A)

$326/283 = (12/24)^b$ (take b outside of the brackets)

$1.152 = (.5)^b$ (Simplify)

Log 1.152 = b*Log(.5)

0.0614 = b(-.301)

$.0614/-.301$ = b

$-.204$ = b

Log (LC) / Log (2) = b = -.204

Log (LC) = -0.204*0.301 = -0.0614

LC = $10^{-0.0614}$ = 86.8%

Now you can plug this information back into the classic learning curve equation to solve for A.

$Y = Ax^b$ therefore

$326 = A(12^{-0.204})$

$A = \$541.26$

Now let's prove that A is in fact $541.26 by demonstrating that it provides the same answer as the known value for unit 24.

$Y_{24} = \$541.26*(24^{-0.204}) = \$283$

Now we have finished with the preliminaries. The information contained in this chapter is necessary for the understanding of learning curves. Be warned however that what we have seen so far are the simple applications. I'm afraid that much more detailed math lies ahead. However, I doubt you would have needed this text if you only had simple questions to answer. Stay with me as the road ahead gets bumpy.

Chapter 2: Single lot learning curves

Suppose I have the following data:

Lot 1 Cost $100,000
Lot 1 units 300

What is my unit learning curve?

The question can't be answered from just this data set. Let me show you why by looking at the standard learning curve equation:

$$Y = Ax^b$$

If I knew the learning curve I could determine a midpoint (more on those later) to determine the value of x. If I knew the midpoint I could determine the learning curve to determine the value of b. I know neither.

So what do I do?

Well the truth is I'm stumped.

I solved for 4 learning curves and found the following:

LC:	A
95%	$471

90%	$675
85%	$978
80%	$1,435

All four values work. When I have to solve lot data and I only have lot information then I need to know another value (such as A) or have at least two data points (as I will show later) to solve. Now there are assumptions that can be made (such as having a knowledge of the range of learning curves, having enough information to make a guess at the A value, etc) that will let you provide a decent estimate but it cannot be as good an estimate as you could get with more information. There just isn't enough data to determine what the actual learning curve values are.

Now let's suppose I have the following information:

My records show that it took $100,000 to make the first lot of 300 widgets and the first unit cost was $705. Now what is the learning curve?

$$Y = Ax^b$$

$$Y = \$705(x^b)$$

Do I have enough information to solve for the learning curve.

The answer is yes. Here's how to do so for a unit learning curve.

$Y_1 = \$705 \, (1^b)$; $Y_2 = \$705 \, (2^b)$; …; $Y_{300} = \$705 \, (300^b)$

Now $Y_1 + Y_2 + … + Y_{300} = \$100,000$

Combine the equations:

$\$100,000 = \$705 \, (1^b) + \$705 \, (2^b) + … + \$705 \, (300^b)$

So I have

$\$100,000/\$705 = 1^b + 2^b + … + 300^b$

$\$141.84 = 1^b + 2^b + … + 300^b$

Now you can solve by brute force (repeated iteration or goal seek) on a computer.

In this particular instance I get a learning curve of 89.4%. You will see in the next chapter how to solve this type of question with a midpoint and will see how much easier the math becomes when midpoints are used.

Chapter 3: Midpoints

Normally we are not lucky enough to get a question such as:

What is the labor required to build the 4th unit ?

Instead the type of question we have to answer is:

I've got to bid a lot of 300. How many labor hours should I request?

To answer the second question we can either do a summation of 300 individual learning curve calculations of the type Y = $A(x^b)$ or else use a midpoint that algebraically represents the lot.

I'll show a small example to demonstrate how much easier it is to use a midpoint to do our work rather than a calculation.

Assume the following:

A = 400
b = -0.152 (90% learning curve)
We are in the first lot

How much labor should we charge for the 300 unit lot?

To determine the labor by summing all the individual learning equations we would have to solve the following 300 times:

$$Y_1 = 400(1^{-.152})$$

$$+ (Y_2 = 400(2^{-.152})) + \dots$$

$$+Y_{300} = 400(300^{-.152})$$

You can do this relatively easy on a computer, but it takes a lot of time and takes a lot of computer memory. Also, if you have to have the data checked, it can eat up a fellow engineer's email memory very quickly and can cause him to lose his normally placid demeanor (this I know from experience).

So what do I do instead?

First calculate the midpoint (take my word for now that this is the midpoint, I'll demonstrate how to calculate it later on in the text):

$$Y_{mid} = A(x_{mid}^{-0.152}) = 400(103.42^{-0.152}) = 197.62$$

Hours for the total lot: 300*197.62 = 59,287

If I do a summation of all of the 300 units
I also get 59,287 (or close to it depending
upon how the rounding is handled by my
computer).

Midpoints are more than just a time saver
however, often all you have is lot data
and from that you have to determine
learning curves, A or some other value.
Midpoints make the math much easier
when dealing with these types of
questions.

Now how do I calculate a midpoint?

Well it is not the average of the first
unit in and the last unit out. That is
important to remember. Following is
the formula and let me apologize in
advance for forcing you to reach for
the headache powder.

$X_{mid} = \{(b+1)*(\text{lot size})/[(\text{lot size} + \text{prior units} +0.5)^{\{b+1\}} - (\text{prior units} +0.5)^{(b+1)})]\}^{-1/b}$

Now quick solve the equation for a
first lot of 101 and a learning curve of
90%.

Just kidding. Let me show you the easy
way to understand the equation and

(more importantly) how to program it into your spreadsheet.

Set up a spreadsheet this way:

Let L = Lot number
Z = Number of units in the Lot
N1 = First unit of the lot
N2 = Last unit of the lot
LC = Learning Curve assumed
b = learning exponent (derived from Learning curve)
1 + b = learning exponent plus 1
-1/b = -1 divided by learning exponent
A= Z(1+b) = Number of units in the lot multiplied by 1+b
$B = (N2 +0.5)^{(1+b)}$= Last unit of the lot + 0.5 all raised to the 1+b power
$C = (N1 +0.5)^{(1+b)}$= First unit of the lot + 0.5 all raised to the 1+b power
D = B-C
E = A/D
Midpoint = $(E)^{(-1/b)}$

Let's see how this plays out with the following:

Lot 1, 90% LC, 300 units
Lot 2, 90% LC, 250 units
Lot 3, 85% LC, 200 units

The results are:

L	Z	N1	N2	LC
1	300	0	300	90.00%
2	250	300	550	90.00%
3	200	550	750	85.00%

b:	1+b	-1/b
-0.152	0.847997	6.578813
-0.152	0.847997	6.578813
-0.23447	0.765535	4.265024

B= (n2+.05) ^(1+b)	C= (n1+.05) ^(1+b)	D = B-C	E = A/D	Midpoint = E^(-1/b)
126.24	0.555556	125.69	2.024075	103.42
210.94	126.2422	84.70	2.503081	418.30
158.92	125.356	33.56	4.561521	647.32

Let's follow one of the calculations through to see how it works:

$L = 1$
$Z_1 = 300$
N1 = 0 (there are no preceding units)
N2 = 300 (the last unit in the lot)
LC = 90% = .90
$b = \log(.90)/\log(2) = -0.152$
$1+b = 1-.152 = 0.848$
$-1/b = -1/-0.152 = 6.579$
$A = Z(1+b) = 300(0.848) = 254.39$
$B = (N2 + 0.5)^{(1+b)} = 300.5^{0.848} = 126.24$
$C = (N1 + 0.5)^{(1+b)} = 0.5^{0.848} = 0.55$

$D = B-C = 126.24-0.55 = 125.69$

$E = A/D = 254.39/125.69 = 2.02$

Midpoint $= E^{(-1/b)} = 2.02^{6.58} = 103.42$

Now let's look at the second calculation

$L = 2$

$Z_2 = 250$

$N1 = 300$ (the N2 value of the last lot)

$N2 = 550$ $(Z_2 + N1)$

$LC = 90\% = .90$

$b = \log (.90)/ \log (2) = -0.152$

$1+b = 1-.152 = 0.848$

$-1/b = -1/-0.152 = 6.579$

$A = Z (1+b) = 250 (0.848) = 211.99$

$B = (N2 + 0.5)^{(1+b)} = 550.5^{0.848} = 210.94$

$C = (N1 + 0.5)^{(1+b)} = 300.5^{0.848} = 126.24$

$D = B-C = 210.94-126.24 = 84.70$

$E = A/D = 211.99/84.70 = 2.50$

Midpoint $= E^{(-1/b)} = 2.50^{6.58} = 418.30$

Notice that the learning curve changed for the third lot. This change will not affect the Z, N1 or N2 values. Please note that the learning curve calculated here is not for all units (i.e. the 85% learning curve applies only to the lot under examination). A calculation would have to be used to determine the learning curve for all 3 lots, I just wanted to show that different learning curve values can be entered into the spreadsheet and it will still calculate the parameters for that lot.

Let's see what happens to the midpoint when we changed learning curves between lots:

$L = 3$
$Z_3 = 200$
$N1 = 550$ (the N2 value of the last lot)
$N2 = 550 + 200 = 750$ $(Z_2 + N1)$
$LC = 85\% = 0.85$
$b = \log (0.85)/ \log (2) = -0.2344$
$1+b = 1-0.2344 = 0.7655$
$-1/b = -1/-0.2344 = 4.265$
$A = Z_3 (1+b) = 200 (0.7655) = 153.11$
$B = (N2 + 0.5)^{(1+b)} = 750.5^{0.7655} = 158.92$
$C = (N1 + 0.5)^{(1+b)} = 550.5^{0.7655} = 125.36$
$D = B-C = 158.92-125.36 = 33.56$
$E = A/D = 153.11/33.56 = 4.56$
Midpoint $= E^{(-1/b)} = 4.56^{4.265} = 647.32$

I recommend you prove out the midpoint calculation the first time you program it. What I mean by this is to take a known value (i.e. 400 lot at a 90% learning curve as I showed at the start of this chapter), sum all of the values in a spreadsheet, then plug the same values into your midpoint calculation. The answers from the two methods should be very close.

Chapter 4 : Learning curves that include production rate adjustments

Some companies use a production rate adjustment factor to the learning curve. The thought behind this is that learning is affected by not only the number of times that a job is done but how often in a time period that job is done.

The equation for a production rate adjusted learning curve is:

$Y = Ax^b r^c$ where

Y = xth unit time
A = time to do the first unit
x = unit under consideration
r = rate per time period (week, month, year etc.)
c = rate exponent = log (rate improvement)/log (2)

Notice that the rate exponent c is calculated just like the learning curve exponent b.

As always let's do an example and see what happens:

Let A = 1,000 hours
Learning curve = 90%

b = log (0.90)/log (2) = -0.152
rate = 100 per month
rate improvement = 99%
c = log (0.99)/log (2) = -0.0044/0.301 = - 0.0145

Let's set up a spreadsheet for the first 5 units for the month:

Unit	Learning Curve	Learning With Rate
1	1,000	935
2	900	842
3	846	792
4	810	758
5	783	732
	4,339	4,059

Note that the addition of a rate adjustment increases the learning by 280 hours (4,339 – 4,059) in this example.

I have not seen any studies that indicate that this phenomena really exists but I have seen a few bids where a company bid this method as part of their payment methodology.

If I am making a comparison between two firms that have bid different methods, I can simply run the two scenarios to see which is better.

For instance:

The Lot the company has to bid is 400
This is a 4 month bid

Firm A bids the following:

85% learning curve
A = 1,000 hours

Firm B bids the following:

90% learning curve, 95% rate adjustment
A = 1,000 hours

Which should I choose?

First calculate Firm A's bid:

85%, 400 lot midpt = 130.77

$b = \log (0.85)/\log (2) = -0.2344$

Firm A bids: $1,000*130.77^{-0.2344} = 319.07$
unit hours

Total hours = 400*319.07 = 127,629

Next calculate Firm B's bid:

90%, 400 lot midpt = 137.35

rate adjustment = 100^c

$b = \log (0.90)/\log (2) = -0.152$

$c = \log (0.95)/\log (2)$

$c = -0.074$

Firm B bids: $1{,}000 * 137.35^{-0.152} * 100^{-0.074}$
$= 336.54$

Total hours $= 336.54 * 400 = 134{,}618$

In this case choose Firm A as it is 7,029 hours less than the bid from firm B.

Is there a learning curve that I can calculate for firm B that will incorporate the rate adjustment into the learning curve?
Let's see:

If I know that A = 1,000 hours I have to calculate the following:

$Y = Ax^b$

$336.54 = 1{,}000\ x^b$

I have one equation and two unknowns so I cannot solve absolutely, but I can do the following:

Calculate the b and the midpt for a lot of 400

LC:	b:	Midpt	Unit prediction
95%	-0.074	143.25	692.54
90%	-0.152	137.35	473.19
85%	-0.2345	130.77	318.97
80%	-0.322	123.42	212.19

We can see that the equivalent learning curve is between 85% and 90%. I can use a spreadsheet and iterate until I have the correct answer, in this case it is: 85.67% with a midpoint of 131.70.

Can this learning curve be used to predict future lots?

Let's say the next lot is for 300 units and let's assume that the bidder will again bid a 90% learning curve and a 95% rate adjustment where A is still equal to 1,000 and r = 100/month:

If I run the analysis as $Y = Ax^b r^c$

I get a total of 81,937 hours

If I run the analysis as $Y = Ax^b$ where the learning curve is 85.67%
I get a total of 73,631. Clearly, we must use the original form to project future estimates.

Chapter 5: Forgetfulness Curves

If a job or process is interrupted for some length of time and then restarted, can we use the old learning curve data to model what the time will be for the restart?

Yes but usually we have to adjust for the break in time:

Assume the following:

A = 1,000 hours
LC = 90%
b = -0.152
Lot = 400
Midpt = 137.35

Y_{unit} = 1,000 * $137.35^{-0.152}$ = 473.2

Now there is a production break of 3 months

Can I start the next lot at unit 401 and continue my projection?

That is doubtful. People forget as well as learn. Studies have shown that forgetting is not complete however, so you don't have to start the learning curve over. How do I handle this?

If you have data of a past occurrence you can use that to estimate what will happen to the time. Let's say that you have noticed that there is a curve correction of 20% when there is a significant break in production. Well let's take this back up the learning curve and see where we should start the next lot:

$Y_{unit\ new}$ = 1,000 * 137.35$^{-0.152}$ = 473.2 *1.20 = 567.84

567.84 = 1,000 *x$^{-0.152}$

Log 567.84 = log 1,000 −0.152 log (x)

2.754 = 3 −0.152 log (x)

-0.246/-0.152 = log (x)

log (x) = 1.618

Antilog (1.618) = x = 41.36

So we should start our new projection from the midpoint 41.36

This midpoint represents a lot of 117.5

For instance, if the new lot is 300 then we would determine the midpoint as 41.36 for the first lot and then the midpoint of the next 300 for the next lot.

What does that do to us?

Assume the following with no break in production:

Lot 1 = 400
LC: 90%
b = -0.152
midpt = 137.35
A = 1,000

Lot 2 = 300
LC: 90%
Midpt = 542.51
A = 1,000

Unit hours for Lot 2 = $1,000*542.51^{-0.152}$
=384.03 hours

Total lot hours = 300 * 384.03 = 115,209

Assume the following with a break in production and a 20% adjustment

Lot 1 = 117.5
LC: 90%
b = -0.152
midpt = 41.36
A = 1,000

Lot 2 = 300
LC: 90%

b = -0.152
midpt = 250.52

$Y_{unit} = 1,000 * 250.52^{-0.152} = 431.9$

Lot total = 300*431.9 = 129,570

The production break has cost us:

129,570 - 115,209 = 14,361 hours

This is one of many ways in which you can handle forgetfulness. The method you use depends upon the data that you have. If there is no data then you are on your own. Just remember that there will be forgetfulness (if you don't believe it open your college physics book and try to solve a few of the elementary problems) and that seldom does the curve go back to unit 1.

Chapter 6: Solving for Lot to Lot Problems

For the most part, the types of problems that you will encounter will not be unit specific but rather lot specific. For instance:

Lot 1 was 210 units and the cost (note I am using cost and not hours but it doesn't affect the learning curve because cost is usually just labor hours time labor rate) is $30,000

Lot 2 was 315 units and the cost was $33,507

What is the learning curve?

Recall the basic learning equation:

$$Y = A\,x^b$$

Can I solve this basic equation with the data I have?

Yes. But to reduce the amount of math I must do I will introduce another equation:

$$Y = \{(A/(1+b))*(x^{1+b}-1)\}/x$$

Where y = average unit cost (or hours)

A = First unit cost (or hours)
b = learning exponent
x = last unit for the lot in question

How does this help us?

First I can calculate y_1 and y_2

$Y_1 = \$30,000/210 = \$142.86/\text{unit}$

$Y_2 = (\$30,000 + \$33,507)/(210 + 315) = \$120.96/\text{unit}$

Note that Y_2 = the sum of both lots

Now I know that x_1 and x_2 are 215 and 525 (215+310)
Respectively

Now let's see what I have

EQ 1: $\$142.86 = \{[A/(1+b)]^*(215^{1+b}-1)\}/215$

and

EQ 2: $\$120.96 = \{[A/(1+b)]^*(525^{1+b}-1)\}/525$

For those of you that recall 10th grade algebra, I have
The classic two equations and two unknowns (A and b are unknown)

These are nonlinear equations so I have to go to logarithms again to solve:

First let's do some substitution:

Let $A/(1+b) = D$

Our equations become:

EQ 1: $\$142.86 = \{D*(210^{1+b}-1)\}/210$

and

EQ 2: $\$120.96 = \{D*(525^{1+b}-1)\}/525$

Now let's remove the divisor

EQ 1: $\$142.86 * 210 = \$30,000 = D(210^{1+b}-1)$

EQ 2: $\$120.96 * 525 = \$63,507 = D(525^{1+b}-1)$

Notice that $D = A/(1+b)$ is the same value for both equations

Let's substitute e for 1+b

EQ 1: $\$30,000 = D(210^{e}-1)$
EQ 2: $\$63,507 = D(525^{e}-1)$

Now let's take the logs of the two equations

Log (30,000) = 4.477 left side of equation 1
Log (63,507) = 4.803 left side of equation 2

Log D + e*log 210-0 right side of equation 1
Log D + e* log 525 −0 right side of equation 2

Log of 210 = 2.322
Log of 525 = 2.720

Notice that I let the 1 value go to 0 and this is not precisely correct mathematically but I'll explain why I can do that later but first stay with me as I continue through the example:

The log transformations are:

EQ 1: 4.477 = Log D + 2.322 e
EQ 2: 4.803 = Log D + 2.720 e

Now subtract EQ 2 from EQ 1:

4.477 − 4.803 = (Log D-Log D) + (2.322 e −2.720 e)

-0.326 = -0.398 e

Solve for e: e = -0.326/-0.398 = 0.819

Recall that we substituted 1+b for e

Therefore: b = e-1
           b = 0.819-1 = -0.181

Recall further that b = log (LC)/ log (2)

Log (LC) = -0.181 * 0.301 = -0.0545

Antilog −0.0545 = $10^{-0.0545}$ = 88.2 %

I have my learning curve so now I can solve for A:

Choose equation 1 in it's transformed form:

EQ 1: 4.477 = Log D + 2.322 e

Now solve for D:

EQ 1: 4.477 = Log D + 2.322 (0.819)
           4.477 = Log D + 1.902
           Log D = 4.477-1.902
           Log D = 2.575
                D = Antilog (2.575) = $10^{2.575}$ =
375.84

Recall that D = A/(1+b)

$$375.84 = A/(1+b) = A/(1-0.181)$$

$$A = 375.84*0.819 = \$307.45$$

Now if I use these values and do a total summation of the lots 210 +

315 I get: Total costs of lots 1+2 = $63,215.32

My original data showed me the total cost was:

Lot 1: $30,000
Lot 2: $33,507
        $63,507

My error for the calculation was:

$$(\$63,507-\$63,215.32)/\$63,507 = 0.0045 = 0.45\%$$

I am off by less than 1%.

Now about allowing 1 to go to 0 in the equation: $\text{Log } D + e* \log (210-1)$

The logarithm I am trying to take is of the entire portion in parenthesis:

$$\text{Log } (210^e -1)$$

I know of no way to take this log and by ignoring the 1 I have introduced error into my estimate. For instance suppose the value for e = 0.82

Log $(210^{0.82} - 1)$ = Log (79.20) = 1.899

When I take the Log in this form: Log (210^{e}) – Log (1) and again let e = 0.82 I get

Log $(210^{.82})$ – Log (1) = Log(80.20) – Log (1) = 1.904
So what is my error?

(1.904-1.899)/1.904 = 0.0026 = 0.26%

I seldom worry about being off by such small percentages so I just use the learning curve and A values that I calculate.

If you want to impress the boss you can take the learning curve and A values and iterate on the computer until you have the exact value. If I keep A at $307.45 my learning curve becomes 88.07%. If I keep my learning curve constant at 88.2% my A value becomes $304.12.

Now you may ask can I use the classic equation to derive my learning curves?

Let's see:

$$Y = A*x^b$$

So $Y_1 = A*x_1^b$
    $Y_2 = A*x_2^b$

In this instance x_1 and x_2 are midpoints

Midpoints are determined by the lot size and the learning curve

I know the lot size but not the learning curve.

However I do know that the learning curve and midpoint are unique and related (i.e. each learning curve has one midpoint and each midpoint has one learning curve for any lot).

Wait though, the Y value has to be for the midpoint and that introduces another unknown. So I have the following unknowns:

X,b and Y and A

Even with X and b being related and really being only one unknown I still have 3 unknowns. Now I can solve by summing all the y values on a computer but I cannot solve for two equations when

I have 3 unknowns. I can probably solve when I have 3 sets of data but as you will see later there are other (and I think better) ways of calculating learning curves for multilot data. The short answer is, the form I have shown is the easiest way to get an accurate answer quickly however you can always brute force learning curves by setting up the appropriate equations in a spreadsheet.

Chapter 7: Learning Curves and Inflation

When you are doing multiyear calculations to determine the learning curve values you must correct for inflation. Let's take the following data set and see how we can determine the learning curve.

In 2001 we bought 400 items at $38,534
In 2002 we bought 700 items at $52,345

What is my learning curve?

Well I can solve the learning curve by the method I showed in chapter 5. But if I do so without adjusting for inflation I will inflate the learning curve. For instance, using the method I showed in chapter 5 I will get the following:

A = $203
LC: 90.01%

But how much of the learning is accounted for by inflation?

Assume our inflation rate for this item for 2001-2002 was 1%. Then to determine the true learning curve I must first remove this amount from the 2002 cost. Why?

Because I am using dollars as a proxy for hours. If I use hours only, then there will be no inflation and therefore my numbers will not be biased. But when I use dollars/cost I must adjust or else what I claim as learning is incorrect and my future projections will also be incorrect.

In this case let's see how to handle the problem:

Inflation rate: 1%

I can use either base year 2001 or base year 2002. Let's choose 2002 because it's easier.

To get my costs into base year 2002 I must multiply my 2001 costs values by 101%.

In this example I now have the following data set:

In 2001 we bought 400 items at $38,534 which is equal to $38,919 in 2002 dollars

In 2002 we bought 700 items at $52,345 which is equal to $52,345 in 2002 dollars

How did this affect our learning curve values?

I calculate an 89.66% Learning Curve and an A of $211

Well that's not to bad but suppose inflation was 4% (I personally have seen inflation rates of 10-12% during the late seventies) then the data set becomes:

First I multiply my 2001 cost by 104%

In 2001 we bought 400 items at $38,534 which is equal to $40,075 in 2002 dollars

In 2002 we bought 700 items at $52,345 which is equal to $52,345 in 2002 dollars

I calculate an 88.64% Learning Curve and an A of $235

If I use my original values of

A = $203
LC: 90.01%

I will overstate my learning curve by 1.5% on the learning curve and by 13% on the A side.

If you have a contractual requirement to adhere to a certain learning curve make certain that you know how inflation will be handled. For a multiyear contract, the impact of inflation will be significant even for small values of 1-2% per year. If you are on the producing side, this is terrible because part of your learning is being eaten by inflation.

If I want I can choose to use 2001 as a basis and deescalate the 2002 costs to 2001. Here's how to do that:

In 2001 we bought 400 items at $38,534
In 2002 we bought 700 items at $52,345

To get the 2002 buy into 2001 dollars I must do the following:

Inflation rate of 1%

$X(1.01) = \$52,345$
$X = \$52,345/1.01$
$X = \$51,826$

This means our data set becomes

In 2001 we bought 400 items at $38,534 in 2001 dollars
In 2002 we bought 700 items at $51,826 in 2001 dollars

Our learning curve will remain about the same at 89.66% however our A value will now be expressed in 2001 dollars and it will go to $209.

Chapter 8: Data Gathering

It saddens me to say this, but this is usually the toughest part of determining an in-house learning curve. Data is normally not gathered in a manner that makes it easy to obtain and use for determination of a learning curve. For instance:

Companies often gather data by project or by item. Why is this a problem? Because learning curves are associated with touch labor and improvement curves are associated with processes. If you have all labor and cost, including supervision, accounting, engineering, finance, production, etc. it is difficult if not impossible to determine what should and should not be used in the learning curve data set.

Let me give an example:

I have the following information:

I have 350 hours for 200 items for time period 1.

I have 400 hours for 300 items for time period 2.

Now I should be happy. I have enough information to solve my learning curve. In this case my learning curve and A values are:

Learning curve: 88.99%
A : 3.55 hours

Looks good right. So I am asked to forecast the next lot of 500 items.

I say fine.

Midpt lot 3 = 733.85

$b = \log(.8899)/\log(2) = -0.1683$

Projection:

$Y = 3.55(733.85^{-0.1683}) = 1.17$

Lot total = 500 * 1.17 = 585 hours

Now we go merrily along our way until the next lot is completed. The billing for the next log comes in and it is 548 hours.

My boss quietly asks me if I am a complete idiot and when do I wish to tender my resignation.

Well this has happened to a lot of engineers and I'll show you why.

The presumed data set was:

Lot	Units	Hours
1	200	350
2	300	400

Here is the actual data set

Lot	Units	Fixed Hours
Variable Hours	Total Hours	
1	200	100
250	350	
2	300	100
300	400	

Now when I solve for this data set I get the following:

A = 2.25 hours
LC = 90.78%

Now I make my projection:

$Y = 2.25(734.26^{-0.1396}) = 2.25(0.398) = 0.896$ hrs/unit

Variable hours = 500*0.896 = 448
Fixed hours = 100
Total lot hours = 448+100 = 548

The error due to the fixed hours being calculated as variable hours is (585-548)/548 = 6.75%

Now how can I avoid this situation?

Well, I've been doing this type of work for a long time and it still trips me up. So all I can tell you is to be very careful about your data set and to make sure you ask questions about how the data has been gathered.

A further note of caution; different labor types have different learning curves. For instance, touch labor usually comes down the learning curve much faster than most other types of labor. Make sure that you collect the data in the format you need for whatever your application happens to be.

What if I am trying to break down a bid from a contractor and he has not provided me with the information as to fixed costs and variable costs? Well, once you have three lots, you can solve three equations and three unknowns and get a pretty good idea. However, be warned, that learning curves are not always nice and pretty and ever decreasing. My experience has shown me that often learning is represented by a swift decline, followed by no or a small decline and then a swift decline again. You may be seeing different learning exponents and not fixed costs. The best way to obtain the information is to insure that the bidder knows he must bid a fixed cost and a variable cost.

Here's an example of how to solve for a fixed cost plus variable cost situation.

Assume the following:

Lot 1, 100 units, $129,769 Total Cost
Lot 2, 200 units, $168,076 Total Cost

Let's solve this and see what our projection would be for a lot 3 of 500 units.

From Chapter 6 recall that:

$$Y = \{(A/(1+b))*(x^{1+b}-1)\}/x$$

W here y = average unit cost (or hours)
A = First unit cost (or hours)
b = learning exponent
x = last unit for the lot in question

So we will solve for the average cost of Y_{100} and Y_{300}

$$\$129,769 = \{[A/(1+b)]*(100^{1+b}-1)\}$$

($129,769 + $168,076) =
$\{[A/(1+b)]*(300^{1+b}-1)\}$

The solution to this data set is:
LC= 84.45%
A = $3,015
b = -0.2438

Now my next lot is 500 so I can do my midpoint calculation to determine what my projection should be:

Midpoint = 525.69

Projection for the 3rd lot = $3,015 *
$525.69^{-0.2438}$ = $654.60

Projection for the total lot cost =
500*$654.60 = $327,300

Now after the third lot has been produced I check the numbers and determine that the real cost was $291,397. This is an error of |$327,300-$291,397| / $291,397 = 12.3%

Once I see this big an error from projection I might be inclined to investigate further.

Let's say that I have investigated and found that the data set does contain fixed and variable cost but no one will tell me the proportion. So what do I do now?

Set up the average unit cost equations with another unknown:

EQ 1: $129,769 = [\{[A/(1+b)]*(100^{1+b}-1)\} + D$

EQ 2: ($129,769 + $168,076) = $[\{[A/(1+b)]*(300^{1+b}-1)\} + 2D$

EQ 3: ($129,769 + $168,076 + $291,397) = $[\{[A/(1+b)]*(800^{1+b}-1)\}] + 3D$

 Note that I have to add a D for each lot. I also assume that D is a fixed quantity. Usually D is not a true fixed cost, but rather a semifixed cost. Each order has various departments and various overheads that attach. These will be slightly different each time, however, it will usually be fairly close to a fixed cost.

You now have 3 equations and 3 unknowns.

As much as I hate to say this, let's solve:

Let's set up a table for our costs:

Lot   Cost   Cumulative
Cost

1        $129,769 $129,769
= Variable Cost 100 + D
2        $168,076 $297,845
= Variable Cost 300 + 2 D
3        $291,397 $589,242
= Variable Cost 800 + 3 D

First let's substitute C for 1+b

EQ 1: $129,769 = [{[A/C]*(100^c-1)} + D

EQ 2: ($129,769 + $168,076) =
[{[A/C]*(300^c-1)} + 2 D

EQ 3: ($129,769 + $168,076 + $291,397)
= [{[A/C]*(800^C-1)}] + 3 D

Move D over to the other side:

EQ 1: $129,769 - D = [{[A/C]*(100^c-1)}

EQ 2: $297,845 - 2 D= [{[A/C]*(300^c-1)}

EQ 3: $589,242 – 3 D = [{[A/C]*(800^C-1)}]

Let A/C = J

EQ 1: $129,769 - D = [J*(100^c-1)]

EQ 2: $297,845 – 2 D= [J*(300^c-1)]

EQ 3: $589,242 - 3 D = [J*(800^C-1)}]

Solve EQ 1 for D

EQ 1: $D = \$129{,}769 - [J*(100^c-1)]$

Substitute into EQ 2

$\$297{,}845 - 2\,(\$129{,}769 - [J*100^c - 1] = [J*(300^c-1)]$

Simplify

$\$297{,}845 - \$259{,}538 + 2\,[J*100^c - 1] = [J*(300^c-1)]$

$\$38{,}307 = [J*(300^c-1)] - 2\,[J*100^c - 1]$

Substitute EQ 1 into EQ 3

$\$589{,}242 - 3(\$129{,}769 - [J*(100^c-1)]) = [J*(800^C-1)\}]$

Simplify

$\$589{,}242 - \$389{,}307 + 3\,[J*(100^c-1)]) = [J*(800^C-1)\}]$

$\$199{,}935 = [J*(800^C-1)\}] - 3\,[J*(100^c-1)])$

Let's solve Eq 2 for J

$\$38{,}307 = [J*(300^c-1)] - 2\,[J*100^c - 1]$

Let's take J out of the right hand side equation:

$38,307 = J[(300^c-1)] - 2 [100^c -1]

Solve for J:

J = $38,307 / [(300^c-1)] - 2 [100^c -1]

Now let's substitute this into Eq 3

$199,935 = [J*(800^C-1)}] - 3 [J*(100^c-1)])

$199,935 =[{$38,307 / [(300^c-1)] - 2 [100^c -1]} *(800^C-1)]

-   3 * [$38,307 / [(300^c-1)] - 2 [100^c -1] *(100^c-1)]

Simplify

$199,935 = $38,307 *(800^C-1)/ [(300^c-1) - 2 (100^c -1)]

- 3($38,307)* [100^c-1]/ [(300^c-1) - 2 (100^c -1)]

$199,935 * [(300^c-1) - 2 (100^c -1)] = $38,307 *(800^C-1) – 3(($38,307)* [100^c-1]

$199,935 * [(300^c-1) - 2 (100^c -1)] = $38,307 [(800^c-1) – 3 (100^c -1)]

[(300^c-1) - 2 (100^c -1)] /[(800^c-1) – 3 (100^c -1)] = $38,307/$199,935

My math skills have about expired so I won't attempt a closed form solution, however, I can take this equation and set up a spreadsheet and solve for c either with iterations or with a goal seek.

My solution yields c = 0.83

b = 0.83-1 = -0.17

LC = 88.9%
A = $1,362

D = $56,418

Now the true values are:

c = 0.848

b = -0.152

LC = 90%

A = $1,200

D = $60,000

So I don't have a completely accurate answer but I do have a much better answer than the one I had before. For instance, let's project a fourth lot based on the information I have now and compare to the way I would have forecasted earlier:

Below is our data set

Lot	Units	Lot Cost
1	100	$129,769
2	200	$168,076
3	500	$291,397

Now I am going to determine my forecast for a fourth lot of 400 units without accounting for a fixed versus variable cost break out. I determine my learning curve to be:

LC: 82.80%
b = -0.272
A: $3,310

I did this by combining lots 2 and 3 and using the method I described in chapter 6 to solve.

If I forecast the next 400 units based on these values I get:

Midpt 82.80% = 991.94

Lot 4 unit projection: $3,310 * (991.94$^{-0.272}$) = $506.74

Total lot projection 400 units * $506.74 = $202,695

The lot 4 projection based upon a fixed cost and a variable cost scenario would be:

Midpt 88.9% = 992.63

Lot 4 unit projection: $1,362 * (992.63$^{-0.170}$) = $421.35

Lot 4 variable projection: 400 * $421.35 = $168,540

Lot 4 total cost: $168,450 + $56,418 = $224,958

The actual cost following a 90% learning curve variable and a $60,000 fixed is:

Variable Cost: $168,156
Fixed Cost: $ 60,000
Total Cost: $228,156

Our errors are:

Variable Cost only projection: ($228,156 - $202,695) / $228,156 = 11.15%

Variable Cost plus Fixed Cost projection: ($228,156 - $224,958) / $228,156 = 1.40%

Now which estimate would you rather defend?

There are other problems that crop up in data gathering, for instance, often costs are gathered by phases. Typically, there is a Design Phase, a Prove out phase, a Production Phase, and (sometimes) a Disposal phase. The trouble with the Design Phase and Prove out Phase is that items are often produced to show proof of concept or for testing and then these parts are used in the production run. If you take the costs that go into developing these very special parts and try to develop an A value then your cost projections will be very high. Also, if 50 or 60 parts are developed during this phase, the people purchasing follow on lots often want you to include those numbers in your learning curve. Don't. These items were not part of the production run. These items are usually built by design engineers and/or especially trained line workers. The learning curve has not yet started to operate and you should not include these items as part of your anticipated costs.

Chapter 9 Step Downs

A step down is defined as lowering the A value used in the learning curve analysis. For instance assume the following:

I have a first lot of 100 items and I have a reliable estimate of $1,000 per item for the lot.

The lot will operate to a 90% learning curve

My A value is: $1,000/35.39^{-0.152} = $1,720

My boss says that we have to sell these things for $800 per item for this lot and that we have to continue to sell these things according to a 90% learning curve for future lots.

How do I handle this?

Well, my new A value becomes $800/35.39^{-0.152} = $1,375

I have a step down of ($1,720 - $1,375)/$1,720 = 20.0%

Now in this case, I could have easily taken the $1,000 and the $800 per item and determined that we would have a 20% step down. But often, you are trying to reach an Annual Unit Cost (AUC) of a certain value and you have to adjust the A value to obtain this AUC. Then you have to determine the step down for comparison purposes.

The important thing to take from this discussion is that when people talk about a step down, it is almost always based upon a difference in A values. So if you are ever asked to take a step down of 20% you will be able to do so with ease.

Does a 20% step down mean that the overall cost will also be a 20% step down? Let's investigate:

Let's take 3 lots and ignore inflation for the moment:

LC: 90%
Lot 1: 100 units, $800/unit
Lot 2: 200 units
Lot 3: 300 units
A = $1,720

Unit forecast Lot 2= $1,720 * 190.30$^{-0.152}$
= $774.56

Unit forecast Lot 3 = $1,720 * 440.65^{-0.152}$
= $681.75

Total cost:

Lot 1: 100 units * $1,000.00/unit =
$100,000
Lot 2: 200 units * $774.56/unit =
$154,912
Lot 3: 300 units * $681.75/unit =
$204,525
Total
$459,437

Now the step down:

LC: 90%
Lot 1: 100 units, $1,000/unit
Lot 2: 200 units
Lot 3: 300 units
A = $1,720
Unit forecast Lot 2= $1,375 * 190.30^{-0.152}$
= $619.20

Unit forecast Lot 3 = $1,375 * 440.65^{-0.152}$
= $545.00

Total cost:

Lot 1: 100 units * $800.00/unit = $
80,000

Lot 2: 200 units * $619.20/unit = $123,840
Lot 3: 300 units * $545.00/unit = $163,500
Total
$367,340

Step down for all lots = ($459,437 - $367,340) / $459,437 = 20%

So yes, it does apply to all the lots.

A step down is just an easy way of handling an overall cost reduction. By applying the savings to the A (or T1) value, your spreadsheet will instantly apply the saving throughout all lots.

Chapter 10 Multilot learning curve analysis

One of the most difficult analyses concerning learning curves is when there are multiple lots. The most used method is to set up a nonlinear regression model and to determine the best fit to a curve. Let me demonstrate how to determine the estimates for a cumulative average curve.

First off let me apologize for the mess you are about to encounter, but that's life.

The standard learning equation: $Y = A x^b$ is a power function.

If you are smart enough to derive the normal equations from this then you shouldn't have to read this book. I can't but I can take them from any standard non-linear statistics book and show them to you.

Normal equations for a power function are

$\text{Log } A = [\Sigma(\log x)^2 \, \Sigma \log y - \Sigma \log x \, (\Sigma \log x \log y)] /$
$$[n \, \Sigma (\log x)^2 - (\Sigma \log x)^2]$$

$$b = [n\Sigma (\log x \log y) - (\Sigma \log x \, \Sigma \log y)] / [n \, \Sigma (\log x)^2 - \{\Sigma \log x)^2\}]$$

Now I know this looks like a bear but it is surprisingly easy to set up a spreadsheet to solve these equations.

Let's define the terms first:

A = First unit cost
b = learning exponent
y = average cost for all units through the xth unit
x = unit under evaluation
n = total number of samples

As always, let's do an example. Here is a table I've set up.

Multilot non-linear Regression

Lot	x	Cum X	Y	Log X
1	300	300	$4,202	2.477
2	400	700	$3,694	2.845
3	300	1,000	$3,499	3.000
4	400	1,400	$3,325	3.146
5	500	1,900	$3,174	3.279
				14.747

Log Y	Log X Log Y	(logx)^2
3.623	8.976	6.136
3.568	10.150	8.095
3.544	10.632	9.000

3.522	11.080	9.898
3.502	11.481	10.750
17.758	52.319	43.879

I will try to simplify the normal equations to make this discussion easier to follow.

Let's see what we have:

$\sum (\log x)^2 = 43.88$

$\sum \log y \quad = 17.76$

$\sum \log x \quad = 14.75$

$\sum \log x \log y = 52.32$

$(\sum \log x)^2 = 14.75^2 = 217.48$

$n = 5$

Log A = [(43.88)*(17.76)-
(14,75*52.32)]/[(5*43.879)-217.48)]

Log A = (779.2217 −771.54484)/1.92

Log A = 4.00

$A = 10^4 = \$10,000$

b= [(5*52.32)-(14.75*17.76)]/ [(5*43.879)-
217.48)]

b = (261.5933-261.8848)/1.92

b = -0.291589/1.92

b = -0.152

Antilog LC = -0.152*.301 = -0.045752

$LC = 10^{-0.045752} = 90\%$

This is a perfect fit of course because I cheated and figured out my data before I did the example.

There are statistical techniques that can be used to determine the fit of the variables to the curve. The one most used is called the regression coefficient (also called R^2). There are several ways to calculate this value but the easiest I've seen is:

Calculate the predicted value for each of the data points (Y_{pred}).
Calculate the average value for each of the data points (Y_{avg}).
Take the predicted value and subtract the average value and then square the result. Sum for all data points (call this A).
Take the actual value and subtract the average value and then square the result. Sum for all data points (call this B).
Divide A by B.

Here is the R squared for the preceding example.

Cumulative Average Learning Curve R Squared Example

Lot	Cum X	Y Pred	Y Act	A = Y Pred - Y Avg	A Squared
1	300	$4,202	$4,202	$623	388,336
2	700	$3,694	$3,694	$115	13,310
3	1,000	$3,499	$3,499	-$80	6,335
4	1,400	$3,325	$3,325	-$254	64,550

5	1,900	$3,174	$3,174	-$405	163,925
			$3,579		636,456

Y Act- Y Avg	B Squared			R Sq	1
$623	388,336				
$115	13,310				
-$80	6,335				
-$254	64,550				
-$405	163,925				
	636,456				

Note that it is a perfect fit. I knew this before I did the calculation because I chose the data to provide a perfect fit.

Here is an example of a non-perfect cumulative average data set:

The data are as follows:

Lot	Units	Cost
1	300	$1,260,640
2	400	$1,557,600
3	300	$ 997,200
4	400	$1,242,800
5	500	$1,500,000

Here is the spreadsheet that I used to determine my A and b values.

Lot	x	Cum x	Y	Log X	Log Y
1	300	300	$4,202	2.477	3.623
2	400	700	$3,894	2.845	3.590
3	300	1,000	$3,324	3.000	3.522
4	400	1,400	$3,107	3.146	3.492

5	500	1,900	$3,000	3.279	3.477
Log X Log Y	(logx)^2			14.747	17.705
8.976	6.136				
10.215	8.095				
10.565	9.000				
10.987	9.898				
11.401	10.750				
52.144	43.879				

If you work through the numbers you should get an A value of $13,270 and a b value of –0.1973.

The R^2 calculation looks like this:

Cumulative Average Learning Curve R Squared Example

Lot	Cum X	Y Pred	Y Act	Y Pred - Y Avg	A = A Squared
1	300	$4,307	$4,202	$802	642,524
2	700	$3,644	$3,894	$139	19,204
3	1,000	$3,396	$3,324	-$109	11,886
4	1,400	$3,178	$3,107	-$327	107,033
5	1,900	$2,992	$3,000	-$513	263,155
Y avg			$3,505.43		1,043,803

Y Act- Y Avg	B Squared
$697	485,400
$389	150,989
-$181	32,916
-$398	158,744
-$505	255,456
	1,083,505

R Sq	0.963358

Again I get a very good R^2 (.9633 or 96.33%).

Now let's work one for a unit curve.

For the unit I have a problem. What is my x value? Is it the last unit. Is it the midpoint? If it is the midpoint, how can I determine a midpoint without first knowing the learning curve? Let's work a 90% learning curve utilizing midpoints and see what we get.

Here is our data set:

Assume the following:
A = $10,000
LC: 90.00%
b: -0.152

Lot	Units	Midpoint	Unit Cost	Lot Cost
1	300	103.42	$4,941	$1,482,160
2	400	484.55	$3,907	$1,562,717
3	300	845.38	$3,590	$1,076,962
4	400	1,194.06	$3,406	$1,362,519
5	500	1,643.19	$3,245	$1,622,466

Let's see what my spreadsheet for A and b looks like:

Lot	Unit learning curve x midpt	Y	Log X	Log Y
1	103	$4,941	2.015	3.694

77

2	485	$3,907	2.685	3.592
3	845	$3,590	2.927	3.555
4	1,194	$3,406	3.077	3.532
5	1,643	$3,245	3.216	3.511
			13.920	17.884

Log X Log Y	(logx)^2
7.441	4.059
9.645	7.211
10.406	8.568
10.869	9.468
11.291	10.341
49.653	39.646

Using this data I do get an A value of $10,000 and a b value of –0.152. It is a perfect match (naturally as I pre-selected the data, I wish I could pre-select the data for all engineering problems I have to work).

What does my R^2 look like?

Unit Learning Curve R Squared Example

Lot	Cum X	Y Pred	Y Act	A = Y Pred - Y Avg
1	103	$4,941	$4,941	$1,123
2	485	$3,907	$3,907	$89
3	845	$3,590	$3,590	-$228
4	1,194	$3,406	$3,406	-$411
5	1,643	$3,245	$3,245	-$573
Y avg		$3,818		

A Squared	B = Y Act- Y Avg	B Squared	R Sq	1

1,260,789	$1,123	1,260,789
7,940	$89	7,940
51,899	-$228	51,899
169,240	-$411	169,240
328,047	-$573	328,047
1,817,915		1,817,915

Okay so now I know that this method works when I know the learning curve ahead of time. But if I knew it ahead of time I wouldn't have to do all this work to determine the blasted thing to start with. Since I don't know what the true midpoint is I am going to select a midpoint (to start with 85%) and see what happens.

Lot	Units	Midpoint X	Y	Log X	Log Y
1	300	99	$4,977	1.994	3.697
2	400	483	$3,908	2.684	3.592
3	300	845	$3,590	2.927	3.555
4	400	1,194	$3,406	3.077	3.532
5	500	1,643	$3,245	3.216	3.511
				13.897	17.888

Log X Log Y	(logx)^2
7.371	3.975
9.642	7.205
10.405	8.567
10.868	9.467
11.291	10.340
49.577	39.554

Let's see what our A and b values are:

A = $10,000
b=-0.152

Changing the midpoint had absolutely no effect (the reason for this is that there is very little difference in midpoints after the first lot as you will see in a later chapter called do midpoints matter). Since that is the case I will proceed cautiously ahead with an unknown data set. Here's the data.

Lot	Units	Lot Cost	Unit Cost
1	300	$1,500,900	$5,003
2	400	$1,781,600	$4,454
3	300	$1,172,400	$3,908
4	400	$1,425,200	$3,563
5	500	$1,500,000	$3,000
	1900	$7,380,100	

Here is the derivation of my A and b values:

Multilot non-linear Regression Unit learning curve
A: $11,188
b: -0.1637

Lot	Units	Midpoint X	Y		Log X	Log Y
1	300	99	$5,003		1.994	3.699
2	400	483	$4,454		2.684	3.649
3	300	845	$3,908		2.927	3.592
4	400	1,194	$3,563		3.077	3.552
5	500	1,643	$3,000		3.216	3.477
					13.897	17.969

Log X Log Y	$(logx)^2$
7.375	3.975
9.794	7.205
10.513	8.567
10.928	9.467
11.181	10.340

49.792   39.554

Here is my R² calculation:

Unit Learning Curve R Squared Example

Lot	Cum X	Y Pred	Y Act	A = Y Pred - Y Avg	A Squared
1	99	$5,277	$5,003	$1,292	1,667,973
2	483	$4,068	$4,454	$82	6,747
3	845	$3,712	$3,908	-$273	74,685
4	1,194	$3,508	$3,563	-$477	227,856
5	1,643	$3,330	$3,000	-$656	430,379
Y avg			$3,985.60		2,407,640

Y Act- Y Avg	B Squared		R Sq	0.998805
$1,017	1,035,103			
$468	219,399			
-$78	6,022			
-$423	178,591			
-$986	971,407			
	2,410,521			

Now I want to do an example where the numbers are all over the place. Here's my data set:

Lot	Units	Lot Cost	Unit Cost
1	300	$1,800,000	$6,000
2	400	$1,200,000	$3,000
3	300	$1,950,000	$6,500
4	400	$1,200,000	$3,000

5	500	$2,200,000	$4,400
	1900	$8,350,000	

I don't see any obvious learning here.
The numbers are not steadily decreasing.
So what does my calculations and R
squared show?

Multilot non-linear Regression Unit learning curve

A: $9,575
b: -0.12354

		Midpoint			
Lot	Units	x	Y	Log X	Log Y
1	300	99	$6,000	1.994	3.778
2	400	483	$3,000	2.684	3.477
3	300	845	$6,500	2.927	3.813
4	400	1,194	$3,000	3.077	3.477
5	500	1,643	$4,400	3.216	3.643
				13.897	18.189

Log X Log Y	(logx)^2	
7.533	3.975	
	9.334	7.205
11.160	8.567	18.19
10.699	9.467	
11.716	10.340	13.90
	50.441	39.554

a	Log A= [(a*b)-(c*d)]/e(a-f)
b	719.4354
	18.45
c	700.9878
d	Log A = 3.98
	A = $9,575

n		5 e	4.63
Sum (Log x)^2		39.55 a	

(Sum log x)^2			193.14 f	

$b = [(e*d)-(c*b)]/e(a)-f$ 252.2025   -0.57247

252.775

4.63

b =	-0.123542
Antilog LC	-0.037186
LC:	91.79%

And my R² is:

Unit Learning Curve R Squared Example

Lot	Cum X	Y Pred	Y Act	A = Y Pred - Y Avg	A Squared
1	99	$5,430	$6,000	$850	722,958
2	483	$4,462	$3,000	-$118	13,974
3	845	$4,164	$6,500	-$416	172,811
4	1,194	$3,990	$3,000	-$590	347,686
5	1,643	$3,836	$4,400	-$744	553,559
Y avg			$4,580.00		1,810,988

R Sq   0.168809

B = Y Act- Y Avg	B Squared
$1,420	2,016,400
-$1,580	2,496,400
$1,920	3,686,400
-$1,580	2,496,400
-$180	32,400
	10,728,000

As you can see my R^2 has fallen tremendously. I do not have any confidence that learning is actually taking place when I get unit prices that vary this wildly.

There is another method that I use to get a quick and dirty answer for unit multilot learning. What I do is take the first lot information, then take the average of all lots for the costs and the total number of information and use logarithms to solve for A and b. Let's see how this works.

Our data set is:

Lot	Units	Lot Cost	Unit Cost
1	300	$1,482,160	$4,941
2	400	$1,562,717	$3,907
3	300	$1,076,962	$3,590
4	400	$1,362,519	$3,406
5	500	$1,622,466	$3,245
	1,900	$7,106,823	
	Avg	$3,740.43	

So I have the following:

Lot 1, 300 units, $4,941 unit cost
All lots, 1,900 units, $3,740 unit cost

Then I solve this via logarithms (using the method I showed earlier based on average cost) to get A and b. For this example I get:

A: $1,014
LC: 90.04 %
b: -0.1512

Since the data represents a pure 90% unit learning curve I have gotten a very good fit indeed. I have compared the nonlinear regression method with this method many times over and I have found that this method usually gives a better answer.   This method is called the multilot average cost method.

Here is the step by step solution if you wish to see it:

$$Y = \{(A/(1+b))*(x^{1+b}-1)\}/x$$

Where y = average unit cost (or hours)
A = First unit cost (or hours)
b = learning exponent
x = last unit for the lot in question

$Y_1 = \$4,941$, $x_1 = 300$ units : $Y_1*x_1 = \$1,482,160$
$Y_{avg} = \$3,740$ $x_{all}=1,900$ units:
$Y_{avg}*x_{all}=\$7,106,823$

We will let $A/(1+b) = D$
And $(1+b) =c$

So we have:

$1,482,160 = D(300^c)$
$7,106,823 = D(1,600^c)$

Time to take our logs and eliminate D:

Log ($1,482,160) = Log D + c * Log (300)
Log ($7,106,823) = Log D + c * Log (1,600)

 6.1709 = Log D + 2.477 c
-6.8517 = - Log D - 3.279 c

Solve for c
-0.6808 = -0.802 c

c = 0.8488

b = c-1 = -0.1512

Anti log LC = -0.1512 * 0.301 = -0.0455

$LC = 10^{-0.0455} = 90.05\%$

Now to solve for A:

6.1709 = Log D + 2.477 c

6.1709 = Log D + 2.477 (0.8488)

6.1709 = Log D + 2.1025

Log D = 2.935

$D = 10^{2.935} = 861.05$

A = D/(1+b) = 861.05/0.8488 = $1,014

As you can easily see this is a very good fit and using my handy dandy spreadsheet it took much less time than the nonlinear regression model.

I will show you how to compare the effectiveness of the two different methods. Let's assume the following data set:

Lot	Units	Cost
1	400	$4,800,000
2	300	$1,405,650
3	500	$2,022,567
4	400	$2,022,567
5	600	$2,022,567

Now let's determine the nonlinear regression A and LC values and R^2 and the average cost method A and LC values.

For the first lot I get an A value of $22,990, a LC value of 91.28% and an R^2 of 95.53%

For the average cost method I get an A value of $21,748 and a LC value of 91.99%.

Which is the better estimate?

To compare the two I use the A value and LC value of both and compare the final estimates with the actual cost. The estimate that comes the closest to the actual costs I judge to be the best. The AD column means absolute deviation and it is the absolute value of the deviation between the estimate and the actual values.

Comparison of nonlinear regression versus average cost method for unit learning curves

Nonlinear values
A: $22,290
LC   91.28%
b: -0.1316

Lot	Units	nonlinear mipoints	Unit nonlinear estimate	nonlinear estimate total
1	400	138.92	$11,643	$4,657,141
2	300	542.66	$9,731	$2,919,358
3	500	937.9	$9,055	$4,527,487
4	400	1,395.09	$8,594	$3,437,544
5	600	1,891.52	$8,256	$4,953,783

Actuals	AD	% Error
$4,800,000	$142,589	2.98%

$3,300,000	$380,642	11.53%
$5,000,000	$472,513	9.45%
$3,600,000	$162,456	4.51%
$4,800,000	$153,783	3.20%
$21,500,000	$1,312,253	6.10%

Lot	Units	mipoints	Unit Average Cost	Average Cost estimate
1	400	139.77	Cost	estimate
2	300	542.73	$10,187	$3,056,026
3	500	938.02	$9,537	$4,768,516
4	400	1,395.14	$9,092	$3,636,694
5	600	1,891.61	$8,764	$5,258,632

Actuals	AD	% Error
$4,800,000	$1,981	0.04%
$3,300,000	$243,974	7.39%
$5,000,000	$231,484	4.63%
$3,600,000	$36,694	1.02%
$4,800,000	$458,632	9.55%
$21,500,000	$972,765	4.52%

For this example the average cost estimate provided a better estimate than did the nonlinear regression model. It has been my experience that it normally does and I use it almost exclusively unless I am doing work for someone that requests the nonlinear method.

Chapter 11 Do Midpoints Matter?

Once when I was really pressed to get an estimate out I used a common midpoint for two estimates. One had 47,000 units the other had 52,000 units. I got caught because my customer did a very simple thing. He divided the total cost of both estimates by the respective units and found out that the unit cost for the 47,000 units was more than the unit cost for the 52,000 units. His question was why? I gave the truthful answer, that there was very little difference between the two and that there was more uncertainty in the estimate than just the few dollars per unit represented by the learning curve. He wasn't totally satisfied with that answer and rightly so. I should have taken the time to do the estimate right. But the truth is there is very little difference in unit cost for a 47,000 unit versus a 52,000 unit estimate based on a fairly steep learning curve. Let's do an example to show what that difference is.

Assume the following:

Estimate one: A = $10,000
            7 lots – 6 lots at 700 and
a seventh lot of 600

LC = 90%

Estimate two: A = $10,000
7 lots – 6 lots at 800 and a seventh lot of 400

LC = 90%

The midpoints are:

Estimate 1:

1	2	3	4
239.02	1,027.51	1,736.94	2,440.86

5	6	7
3,143.01	3,844.38	4,447.80

Estimate 2:

1	2	3
272.88	1,174.23	1,985.00

4	5	6	7
2,789.48	3,591.94	4,393.51	4,998.96

The estimates becomes:

Estimate 1

Lot	1	2	3
Midpt	239.02	1,027.51	1,736.94
Unit Cost	$4,350	$3,485	$3,218
Units	700	700	700
Total	$3,044,869	$2,439,480	$2,252,375

4	5	6	7
2,440.86	3,143.01	3,844.38	4,447.80
$3,056	$2,940	$2,852	$2,789
700	700	700	500
$2,138,854	$2,058,215	$1,996,151	$1,394,571

Total

Unit

4,700 Cost
$15,324,515 $3,261

Estimate 2

Lot	1	2	3
Midpt	272.88	1,174.23	1,985.00
Unit Cost	$4,263	$3,415	$3,153
Units	800	800	800
Total	$3,410,474	$2,731,983	$2,522,436
	4	5	6	7
	2,789.48	3,591.94	4,393.51	4,998.96
	$2,994	$2,881	$2,794	$2,740
	800	800	800	400
	$2,395,299	$2,304,990	$2,235,483	$1,096,021

Total
Unit
5,200 Cost
$16,696,686 $3,211

What I did was use the midpoints for the 52,000 lot for both so my estimate 1 becomes:

Estimate 1 with wrong midpoints

Lot	1	2	3
Midpt	272.88	1,174.23	1,985.00
Unit Cost	$4,263	$3,415	$3,153
Units	700	700	700
Total	$2,984,165	$2,390,485	$2,207,132
	4	5	6	7
	2,789.48	3,591.94	4,393.51	4,998.96
	$2,994	$2,881	$2,794	$2,740
	700	700	700	500
	$2,095,887	$2,016,866	$1,956,048	$1,370,026

Total
Unit

4,700 Cost

$15,020,608 $3,196

What the customer did was to divide the total cost for my estimate $15,020,608 by the number of units 47,000 and he got $3,196/unit.

He also divided $16,696,686/5,200 and got $3,211. Now the total error for this is ($3,261-$3,196)/$3,261 = 1.99%, which is a small error indeed as this was a concept estimate based on mostly parametrics. However, it was a lesson to me that the customer expected certain things in an estimate and that taking a short cut can have ramifications.

Back to my original questions: Do midpoints matter? The answer is maybe. Yeah, I don't like that answer either, but it is the correct one.

When you have very good numbers (production numbers for instance) and you are trying to project future costs for those numbers then you should use the best midpoints you have. If you are in a time crunch and you need to get out a quick and dirty estimate, then you might want to consider using one set of midpoints instead of having to constantly redo midpoints. This is especially true when you have multiple learning curves. Let's look at two common learning curves, 90% and 85% and see what it costs us in accuracy over the lots used in estimate two.

I have:

A = $10,000
7 lots – 6 lots at 800 and a seventh lot of 400
LC1 = 90%
LC 2 = 85%

LC1 midpts (90%):

1	2	3	4
272.88	1,174.23	1,985.00	2,789.48
5	6	7	
3,591.94	4,393.51	4,998.96	

LC 2 midpts (85%):

1	2	3	4
259.33	1,172.32	1,983.89	2,788.69
5	6	7	
3,591.33	4,393.00	4,998.85	

As you can see, in the out years (4 –7) there is hardly any difference in the unit midpoints. As long as you use the appropriate b value there will also be little difference in the estimates. I'll illustrate this with the following:

I will use the 90% midpts with a b of −0.234 and then use the 85% midpoints with a b of −0.234 (this is the b value for an 85% learning curve) and see what difference it makes.

Estimate 1 90% midpts with an 85% b value

Lot	1	2	3	4
Midpt	272.88	1,174.23	1,985.00	2,789.48
Unit Cost	$2,684	$1,907	$1,686	$1,556
Units	800	800	800	800
Total	$2,147,530	$1,525,245	$1,348,589	$1,245,187

5	6	7		
3,591.94	4,393.51	4,998.96		Total
$1,467	$1,399	$1,358		Unit
800	800	400	5,200	Cost
$1,173,516	$1,119,379		$543,002	$9,102,448 $1,750

Estimate 2 85% midpts with an 85% b value

Lot	1	2	3
Midpt	259.33	1,172.32	1,983.89
Unit Cost	$2,717	$1,907	$1,686
Units	800	800	800
Total	$2,173,329	$1,525,827	$1,348,766

4	5	6	7
2,788.69	3,591.33	4,393.00	4,998.85
$1,557	$1,467	$1,399	$1,358
800	800	800	400
$1,245,269	$1,173,562	$1,119,410	$543,005

	Total		
	Unit		
5,200	Cost		
$9,129,168	$1,756		

As you can see the error is almost nothing ($1,756-$1,750)/$1,756 = 0.34%.

As long as the appropriate b value is used then the estimate will be fairly precise. So midpoints matter much less than the b value and you should always be certain that if you do take a short cut that you utilize the appropriate learning exponent.

Chapter 12 Learning Curves and Years

Have you ever heard this one. Let's save a few dollars. Put that production into year 3 instead of year 2. That'll save us a few bucks on the learning curve.

I've heard it and it is dead wrong. Let me show you why.

Let's say I have the following:

6 year cycle
A = $10,000
LC: 90%
Units: 5,000 per year

And let's assume that the fellow making this suggestion wants to slip the schedule to look like this

10 year cycle
A = $10,000
LC: 90%
Units: 3,000 per year

Now for either alternative I get 30,000
units. So my learning will be the same
regardless of the schedule. The estimate
for alternative 1 will be:

Estimate 1 90% learning curve over 6 years for 30,000 units

Lot	1	2	3	4
Midpt	1,605.25	7,324.34	12,396.65	17,426.67
Unit Cost	$3,256	$2,585	$2,387	$2,266
Units	5000	5000	5000	5000
Total	$16,282,369	$12,927,461	$11,933,699	$11,331,628

	5	6	
	22,443.18	27,453.65	
	$2,181	$2,115	
	5000	5000	30,000
	$10,904,147	$10,575,211	$73,954,516

Total
Unit
Cost
$2,465

Estimate 1 90% learning curve over 10 years for 30,000 units

Lot	1	2	3	4
Midpt	964.64	4,394.81	7,438.19	10,456.20
Unit Cost	$3,519	$2,794	$2,579	$2,449
Units	3000	3000	3000	3000
Total	$10,555,736	$8,382,684	$7,738,312	$7,347,913

	5	6
	13,466.11	16,472.39
	$2,357	$2,286
	3000	3000
	$7,070,721	$6,857,427

Lot	7	8	9
Midpt	19,476.73	22,479.90	25,482.33
Unit Cost	$2,228	$2,180	$2,139
Units	3000	3000	3000
Total	$6,685,002	$6,540,863	$6,417,402

	10		
	28,484.25	Total	
	$2,103	Unit	
	3000	30000	Cost
	$6,309,683	$73,905,743	$2,464

As you can see in this example there is a total difference of $1 per unit over the different scenarios and this is not a true difference but rather the difference is due to rounding. **There is no difference in learning due to when the units are put into production.** I have had experienced engineers argue this point with me and if you are in the estimating game very long you will hear the same thing.

There are other good reasons to shift production. One may be that there are budgetary constraints. The customer may want 5,000 units a year but he can only afford 3,000 units per year. Another is that there may be a life cycle comparison. When there is a life cycle comparison then it is always less expensive to have production further out in the life cycle due to the discounting of the values to a base year.

Learning attaches to the unit not to the year. Remembering that can keep you from doing unnecessary runs trying to get the cost down just by shifting the production. As long as you are using constant dollars, it will not make a difference.

If the engineer arguing with you will not go away (a classic engineering trait) then show him the following example:

We have 3 units

A = $10,000
LC: 90%
$b = \text{Log} (0.90)/\text{Log}(2) = 0.152$

The units will be purchased one each in year 1, 2 and 3

The inflation rate is 0% (constant dollars)

The costs are

$Y_1 = \$10,000 \ (1^{-0.152}) = \$10,000$

$Y_2 = \$10,000\ (2^{-0.152}) = \$9,000$

$Y_3 = \$10,000\ (3^{-0.152}) = \$8,462$

Total cost is $\$10,000 + \$9,000 + \$8,462 = \$27,462$

My schedule has now changed and I will be purchasing all 3 items in year three. My costs become:

$Y_1 = \$10,000\ (1^{-0.152}) = \$10,000$ x inflation of 1

$Y_2 = \$10,000\ (2^{-0.152}) = \$9,000$ x inflation of 1

$Y_3 = \$10,000\ (3^{-0.152}) = \$8,462$ x inflation of 1

Total cost is $\$10,000 + \$9,000 + \$8,462 = \$27,462$

There is no change of course.

Chapter 13 When To Go to Standard Hours

Learning attaches theoretically as long as units are being produced but at some point, it makes little sense to continue to use a learning curve. I hope the following example will make my point.

Assume the following:

A = $10,000
10 Lots at 20,000 units per lot
LC 90%

A = $10,000
10 Lots at 30,000 units per lot
LC 90%

Now what if I use the value of the last unit of my 200,000 to estimate the last 100,000 units of the 300,000 estimate.

$Y_{200,000} = \$10,000 * (200,000^{-0.152}) = \$1,564.02$

The cost for units 200,001 – 300,000 =
100,000 * $1,564.02 = $156,402,410

Now what is the cost if I estimate the last 100,000 units using midpts?

The midpt for the last 100,000 units given that 200,000 units have gone before is:
248,062.61

The estimate for the last 100,000 units would be:

$Y_{200,001-300,000} = \$10,000 * (248,062.61^{-0.152}) * 100,000 = \$1,513.65 * 100,000 = \$151,365,000$

What is my error?   ($1,564.02-
$1,513.5)/$1,564.02 = 3.2%

That's a fairly sizeable error so I believe that I must run my learning curve for at least a little while longer. But what about the following?

A = $10,000
10 Lots at 100,000 units per lot
LC 90%

A = $10,000
10 Lots at 105,000 units per lot
LC 90%

My estimated cost for the 1,000,000th unit is:

$Y_{1,000,000} = \$10,000 *(1,000,000^{-0.152}) = \$1,224.62$

Using that to estimate my next 50,000 units I would have
 $1,224.62 * 50,000 = $61,231,000

Using a midpoint of 50,000 beyond 1 million units I get the following estimate:

$Y_{1,000,000-1,050,000} = \$10,000 *(1,024,883.41^{-0.152}) = \$1,220.05$

My estimate for the next 50,000 would be :

$\$1,220.05 * 50,000 = \$61,002,479$

My error is: $(\$1,224.62-\$1,220.05)/\$1,224.62 = 0.37\%$

This I can safely ignore.

So when is the appropriate time to use standard hours. The question to answer is this: When does the accuracy diminish to where it is acceptable to my organization to ignore any future learning. Many mass producers do not use learning curves. They calculate the cost of an item based simply upon the labor, material and mark up. Most of the organizations using learning curves are those that make only a few (where few is loosely defined) of an item. This tends to be organizations in the defense industry where there is only one (or at best a limited number of) customer(s) for fairly expensive goods.

If I am making ink pens and I make 100 Million a year, it is easy to see that the learning curve will not even play into my calculations. What would be the impact on me if I have a very small A value and a very large number of units in a lot:

A= $10
Units: 10,000,000
LC: 90%

My estimated cost per unit with a learning curve is:

Ymidpt = $10 * 3,380,033.18$^{-0.152}$ = $1.02 per unit

Compare this with the way I would estimate this without a learning curve:

It takes 66,667 hours to make 10,000,000 units (based on history of a similar product

My cost per hour is $150/hr

My unit cost becomes: ($150/hr * 66,667 hrs) / 10,000,000 = $1.00 per unit

Why estimate the cost based on total units and not use an A value and a midpoint?

Well, it is usually much easier to come up with an expected number of hours for a production run than to come up with an accurate A value. A is usually a theoretical construct that has been developed after the fact from actual cost and or production data. Why develop this number unless you must have it to project cost estimates. It is usually easier to estimate without learning curves and so learning curves rarely get used in most organizations.

There are cases where learning curves should be used and aren't. One area where learning is rarely used is in construction pricing. The classic construction estimate is based on the size of the structure, the individual complexities of the structure (amount of concrete, size of slab, amount of wood etc.) and the labor is usually estimated based on these factors. The estimators seldom use learning curves for the various disciplines, i.e, an electrician should take less time for the second building he wires than he did for the first. This is not to say that construction estimators are dumb. They often use a weighted average for the whole crew that does include some learning and often the customer will be savvy enough to get the construction Architect Engineer to reduce the bid on a per square foot basis for additional buildings after the first.

My advice, for what it is worth, is to use the learning curve only for the first 100,000 units or so or to when there is smaller than a 1% error on the estimate. Of course, the customer has a large say in this and exactly when standard hours will be used for pricing (and therefore estimates) is often a negotiation point during contract award.

Chapter 14 Making Financial Decision Using Learning Curves

Often an engineer will have to determine whether to change a process, make an insertion, change vendors, etc. Sometimes someone wants to sell you an item to replace an existing item. The price may be higher but it still may be a good deal in the long term.

Here's an example:

I have a widget that has this profile:

A = $600
LC: 90% = b −0.152
Units produced: 2,000

Let's say I need 400 units of these widgets. My cost for this batch becomes:

$600 * Y_{midpt}^{0-152} * 400 = \$600 * 2{,}197.00^{-0.152} * 400 = \$74{,}515$

Another vendor has bid $200 per widget for this lot. Which should we select?

Well on the face of it is very obvious. The second vendor wants $200*400 = $80,000.

We should choose the first vendor as his cost is lower.

Now let's see if the learning curve assumptions would have an impact on our decision.

First let's assume that both have agreed to a 90% learning curve for all of their future work (we will ignore inflation but of course it would be easy to escalate the values if required).

Let's assume we will be ordering the following lots:

Lot A-500 units
Lot B-800 units
Lot C-300 units

In addition to the lot we have currently under consideration. Now which vendor would be the cheapest?

Let's determine the A value for the second vendor. He has agreed to give us 400 units at $200 per unit. He has also agreed to come down a 90% learning curve on the next three lots. Let's compare his costs with the cost for the first vendor.

First let's determine his A value: $A = Y_{avg\ cost} / Y_{mid}^{-b}$

Vendor 2's A = $200/137.35^{-0.152}$ = $423

Please notice two things.

1. A smaller A value.

2. We are farther up the learning curve for this vendor.

These two facts may result in a lower overall price for all four buys even though there is a higher cost for the first buy.

Here are the numbers:

First Vendor

	Lot A	Lot B	Lot C	Lot D
Midpt	2,197.00	2,645.96	3,291.16	3,849.38
Units	400	500	800	300
Unit Cost	$186	$181	$175	$171
Total Cost	$74,515	$90,548	$140,150	$51,319
Totals				

2,000

$356,532
$178

Second Vendor

	Lot A	Lot B	Lot C	Lot D
Midpt	137.35	631.38	1,276.34	1,848.16
Units	400	500	800	300
Unit Cost	$200	$159	$143	$135
Total Cost	$80,000	$79,306	$114,014	$40,416
Totals				

2,000

$313,736
$157

Even though the second vendor is more expensive on the first lot, he is less expensive over all. To do this properly of course, I should set up a discounted cash flow (as I will show in the following example) and determine the present worth of both streams. Since there isn't a large cash outflow here however (no significant first cost for choosing the second vendor) I am very sure that the present worth calculation would show the same winner.

This is a very simplistic scenario of course and often vendors will not commit to a specific learning curve. The point is still relevant however, it is not just today's cost but also the future cost that should be evaluated when comparing vendors.

Another typical scenario is where a vendor will agree to reduce the cost of his widget provided some consideration is given to his firm. To reach the correct financial answer the engineer must know what the tradeoffs are. Usually, there will be a requirement for some up front money (usually called NRE-Non Recurring Engineering in Defense circles) to achieve the desired savings. Learning or improvement curves are often a driver in the determination of whether to invest the up front money in order to achieve the savings.

Let's look at this example:

Proposition 1:
Proposed cost for a widget: $800
Current agreed to learning curve: 90%

Schedule:

Lot 1:   400 units
Lot 2:   800 units
Lot 3: 1,000 units
Lot 4: 2,000 units

If we give the vendor $500,000 today he will reduce the cost for the widgets to $500 for lot 1. He has agreed to come down a 90% learning curve.

Let's put this information into an easy form to use:

Proposition 2:
Proposed cost for a widget: $500
Agreed to learning curve: 90%
Additional funds required: $500,000

To start with let's look at a quick and dirty analysis to see if we want to proceed:

Total units: 4,200
Maximum savings: $300 per unit ($800-$500)
Total possible savings: $1,260,000 (4,200 * $300)

My quick and dirty analysis shows that it may be possible to save some money so I proceed with a proper analysis.

To make this first example easy I will assume no previous units (I will expand the example later to show what would happen if there are units that have already been completed).

First let's determine the two As

Proposition 1: $A = Y_{avg\ cost}/Y_{mid}^{-b} = \$800/137.35^{-0.152} = \$1,691$

Proposition 2: $A = Y_{avg\ cost}/Y_{mid}^{-b} = \$500/137.35^{-0.152} = \$1,057$

Now let's run the numbers to see what the difference is between the two propositions:

First Vendor

	Lot A	Lot B	Lot C	Lot D
Midpt	137.35	759.63	1,671.69	3,139.11
Units	400	800	1,000	2,000
Unit Cost	$800	$617	$547	$497
Total Cost	$320,000	$493,487	$547,163	$994,377

Totals

4,200

$2,355,028
$561

Second Vendor

	Lot A	Lot B	Lot C	Lot D
Midpt	137.35	759.63	1,671.69	3,139.11
Units	400	800	1,000	2,000
Unit Cost	$500	$386	$342	$311
Totals	$200,000	$308,430	$341,977	$621,486

4,200

$1,471,892   Savings $883,135
$350

We see we get a savings of $883,135 based on a first cost of $500,000. To do this analysis properly we will have to determine the return on investment and do that correctly we must have the years that the lots represent as well as the assumed inflation rate.

I will assume the following:

Inflation: 2% per year
Lot A, B, C, D are each 1 year apiece (if there were split years you could take ½ of the units in each year or whatever the appropriate split is)

What is my return on investment (ROI)?

First let's look at the new spreadsheet that includes the escalated costs (inflation adjusted unit prices):

First Vendor with inflation added

	Lot A	Lot B	Lot C
Midpt	137.35	759.63	1,671.69
Units	400	800	1,000
Unit Cost	$800	$617	$547
Esc cost	$800	$629	$569
Total Cost	$320,000	$503,357	$569,269

Lot D		
3,139.11	Totals	
2,000	4,200	
$497		
$528		
$1,055,241	$2,447,867	
$583		

Escalation:	1	1.02	1.0404	1.061208

Second Vendor with inflation added

	Lot A	Lot B	Lot C
Midpt	137.35	759.63	1,671.69
Units	400	800	1,000
Unit Cost	$500	$386	$342
Esc Cost	$500	$393	$356
Total Cost	$200,000	$314,598	$355,793

Escalation:	1	1.02	1.0404

Lot
D

3,139.11	Totals	
2,000		4,200
$311		
$330		
$659,526		$1,529,917
		$364
1.061208		
Savings:		$917,950

What is usually meant by a return on investment is what percentage would my money earn if I made the investment and took the savings? Well most spreadsheets allow you to do this very easily with an at command (such as @IRR(range)). I will show that value then show how you can do it without using the at command. The ROI for this investment is:

Return on investment

Year	Inv	Savings	Cash Flow
0	($500,000)		($500,000)
1		$120,000	$120,000
2		$188,759	$188,759
3		$213,476	$213,476
4		$395,715	$395,715
			23.90%

If I discount all of my yearly savings by 23.90% I should have 0 dollars. Here is that spreadsheet:

Return on investment determined by discounting

118

Year	Inv	Savings	Cash Flow
0	($500,000)		($500,000)
1		$120,000	$120,000
2		$188,759	$188,759
3		$213,476	$213,476
4		$395,715	$395,715
			23.90%

Disc	Present Worth
1	($500,000)
0.8071	$96,855
0.6514	$122,966
0.5258	$112,245
0.4244	$167,934
	$0

The discount is determined this way: Year 0 it is 1 (end of year convention), Year 1 is 1/1.239, Year 2 = $1/(1.239)^2$ etc.

For this example I have a 23.9% return on investment for my $500 K. This is a very good return and I would strongly consider investing the money. However, as always there are many things that can occur and management must weigh everything before making the decision. I personally have seen returns higher than that turned down because of perceived risk and it may have been an appropriate decision

Now what happens if I have prior learning? Well I just adjust my midpoints appropriately and run the numbers. For instance, if I had 2,000 prior units my first midpoint would be 2,197.00 and I would adjust all of my following midpoints accordingly. The preferred candidate would still be the one that had the lowest net present value.

An area where learning curves are often ignored is when a firm does a make versus buy analysis. I worked for years in a manufacturing facility where we did this type of analysis often and not once did I do any sort of a learning curve. Sometimes it doesn't matter (for instance if there are going to be millions of units produced or when the A value is very low) but often it does. Here's an example of how to do an analysis of this type.

Here are the data:

A firm has offered to sell us a better widget for $1,002 per unit for this year and $1,042 next year for lots of 1,000 units this year and 1,200 units next year. Let's call this alternative A.

We have had our engineers look at the possibility of setting up a work cell that would produce the widgets. Their estimates are:

Set up costs for the work cell: $400,000
Material costs: $560 per unit for year 1
and $582 for year 2
Hours to produce: 10 per unit
Our loaded cost per hour is $75 for year 1
and $78 for year 2

Let's call this alternative B.

Let's do a little back of the envelope
analysis to see where we stand.

For alternative A: The costs would be
$1,002 * 1,000 = $1,002,000
For year 1 and $1,042 * 1,200 =
$1,250,400 for a total cost of $2,252,400.

My cost to make the widget is:

$400,000 for the set up
$560 * 1,000 = $560,000 for Material year
1
$582 * 1,200 = $698,400 for Material year
2
10 * $75 * 1,000 = $750,000 for Labor for
year 1
10 * $78 * 1,200 = $936,000 for Labor for
year 2

Total cost for alternative B is $3,344,400

It looks like we should choose alternative A.

But I've done a little bit of investigating and I find out the following:

The work cells are manual work cells and the 10 hours is a rough guess at the time it would take to make the first 10 units. The material vendor does not yet have a contract and these are just estimates by the purchasing department. The purchasing department tells me they based the first estimate on the cost to purchase 100 units and that they used a 4% inflation index to get the cost for the second year.

Historically we have been able to achieve an 85% cumulative average learning curve for manual labor and a 90% cumulative average learning curve for material purchases.

Here is how I would set up my evaluation of alternate B:

First I would determine my A value for the material. From my investigation I find that both the material and labor are based on a cumulative average (i.e. the labor is an estimate of 10 hours per unit for the first 10 units, the material estimate is $560 per unit for the first 100 units). I also know that I have a 90% material learning curve and an 85% labor unit curve.

$$Y = Ax^b$$

$$\$560 = A(100^{-0.152})$$

$$A = \$1,128$$

So my unit material cost for the first 1,000 units are:

$$Y_{1,000} = \$1,128 * 1,000^{-0.152} = \$395$$

And my unit material cost for 2,200 units are:

$$Y_{2,200} = \$1,128 * 2,200^{-0.152} = \$350$$

Now in order to properly apply inflation I have to determine the costs of the first 1,000 units and then the cost of the second 1,200 units.

Lot 1 = 1,000 units * $395/unit = $395,000

Lot (1 + 2) = 2,200 units * $350/unit = $770,000

Lot 2 = Lot (1+2) – Lot 1 = $770,000 - $395,000 = $375,000

Cost for Lot 2 with 4% escalation = $375,000 * 1.04 = $390,000
Now I have to do the same calculation for labor using an 85% learning curve.

$$Y = A*x^b$$

$$10 = A * (10^{-0.2345}) = 17.16 \text{ hours}$$

Hours required for the first 1,000 units:

$$Y_{1,000} = 17.16 * (1,000^{-0.2345}) = 3.40$$
hours/unit

$$Y_{2,200} = 17.16 * (2,200^{-0.2345}) = 2.82$$
hours/unit

The labor cost for the first 1,000 units becomes:

1,000 * 3.40 * $75 = $255,000

The labor cost for the next 2,200 units is calculated by first determining the number of hours in the second lot.

Lot 2 = Lot (1+2) – Lot 1

Lot 2 = (2,200 * 2.82) – (1,000 * 3.40) = 6,204 – 3,400 = 2,804 hours

Lot 2 cost = 2,804 hours * $78/hour = $218,712

Now let's look at the cost of the new alternative B:

Set up cost : $400,000
Material Cost lot 1: $395,000
Material Cost lot 2: $390,000
Labor Cost lot 1: $255,000
Labor Cost lot 2: $218,712

Total Cost: $1,658,712

Based on these set of assumptions alternative B is the winner.

The reason I wanted to show this analysis was to show the reader that there are many places where learning curves should be used and aren't. In my opinion, many mass production firms totally miss the benefits of estimating with learning curves and from that miss an opportunity to at least properly evaluate the potential costs and benefits of making an item versus purchasing an item. There are, of course, many factors that affect the make versus buy decision and most are not economic (for instance, I have to find skilled workers and make sure that I can provide work for them after the two lots have been produced). Still management should have the best economic information available to them when they have to make a decision.

I also have my internal cost to purchase the items, receive the item, store the item etc. but I am going to assume these costs are about the same as the costs to do the same or similar things if I make the item.

Chapter 15 Phillip Marlowe and the Learning Curve

The title of this section may seem strange to those of you under 40 so let me explain a bit. Phillip Marlowe is the classic detective created by Raymond Chandler in the 1930s. The cases he solved were very complicated and yet at the end a logical solution was determined. Learning curves are much the same. In this section I will show you how you can look at different learning curve methods and from that determine what you feel is the best match.

Let's assume the following data set for our widgets:

Year	Units	Cost
1	300	$900,000
2	400	$927,000

Our unit cost becomes: $3,000 per unit for year one and $2,317.50 for year two. Looks like a learning curve right?

Well using logarithms to solve I get the following:

The solution set is:

A: $6,401
LC: 89.23%

This looks reasonable and I may be onto a solution. But I remember that that there might be escalation in the data also. I check the data source and find out that indeed this are inflated dollars. Our current estimate of inflation is 3% so I deescalate year 2 and run the analysis again.

My new data set is:

Year	Units	Cost	Unit Cost
1	300	$900,000	$3,000
2	400	$900,000	$2,250

I see that years 1 and 2 are the same cost. Is this by accident?

Well I run the data set through my tools and find the following:

A: $6,928
LC: 88.15%

I use this to forecast the next lot of 600 units:

$$Y = A * x_{mid}{}^b * units$$

A	Midpt	LC	b:	Y
$6,928	982.4	88.15%	-0.182	$1,977

Units	Constant Lot Cost	Escalated Lot Cost
600	$1,186,478	$1,258,735

My projection is $1,258,735 for the next 600 units and a unit cost of $1,977.

The bid comes in at $954,810 and a unit cost of $1,591. My error is ($1.977-$1,591)/ $1,977 = 19.52%

I am not too happy with my projection. So what is happening? Has there been a change in the learning curve?

Let's run the three data points through the nonlinear regression model and the average cost methods.

First let's look at the data set in constant year dollars:

Year	Units	Cost	Unit Cost
1	300	$900,000	$3,000
2	400	$900,000	$2,250
3	600	$900,000	$1,500

I see that all three years have the same dollar value in constant dollars. That is interesting and I doubt it is a coincidence but I perform the analysis to convince myself that my intuition is correct.

The nonlinear regression model provides the following solution:

Non Linear Regression Equations

Year	Lot	Units	Cost	Unit Cost
1	1	300	$900,000	$3,000
2	2	400	$900,000	$2,250
3	3	600	$900,000	$1,500

Deesc	Constant Dollars	Total
100.00%		
100.00%	$3,000	$900,000
100.00%	$2,250	$900,000

$1,500 $900,000

Lot	Midpt X	Y	Log X	Log Y
1	95.94	$3,000	1.982	3.477
2	482.78	$2,250	2.684	3.352
3	980.92	$1,500	2.992	3.176
			7.657	10.005

Log X Log Y	(log x)^2
6.892	3.928
8.996	7.203
9.502	8.950
25.390	20.081

LC: 82.50%
A: $11,051
R^2: 99.8%

The average cost method gives the following solution:

	Units	Total Cost
Lot 1	300	900,000
Average	1,000	1,800,000

Learning Curve: 84.04%
A: $9,396

The regression coefficent for the nonlinear regression (R^2) is very good (anything over 90% is usually thought of as excellent) so I use this information to determine what my absolute deviation is for the three lots.

I will also use the average cost method to determine what my absolute deviation is for the three lots.

Finally I will forecast a straight fixed cost to see what the absolute deviation is for my hunch that there is really not any learning going on.

Absolute Deviation Comparison

Nonlinear regression calculation

A:	$11,051	
LC:	82.50%	
RSQ:	99.80%	

Lot	Actual	Predicted	Units	Total Actual
1	$3,000	$3,115	300	$900,000
2	$2,250	$1,989	400	$900,000
3	$1,500	$1,634	600	$900,000
				$2,700,000

Total Predicted	Total Error
$934,441	$34,441
$795,720	$104,280
$980,428	$80,428
	$219,149
Error	8.12% AD

Average Cost Extrapolation calculation

A:	$9,496
LC:	84.04%

Total

Lot	Actual	Predicted	Units	Actual
1	$3,000	$2,978	300	$900,000
2	$2,250	$1,994	400	$900,000
3	$1,500	$1,669	600	$900,000
				$2,700,000

Total Predicted	Total Error
$893,389	$6,611
$797,432	$102,568
$1,001,363	$101,363
	$210,543
Error	7.80% AD

Fixed Cost Calculation

Lot	Actual	Predicted	Units	Total Actual
1	$3,000	$3,000	300	$900,000
2	$2,250	$2,250	400	$900,000
3	$1,500	$1,500	600	$900,000
				$2,700,000

Total Predicted	Total Error
$900,000	$0
$900,000	$0
$900,000	$0
	$0
Error	0.00% AD

It is easily seen that the best estimate is for a fixed cost calculation. You might say (and rightfully so) that I should have known that the contractor was bidding a fixed cost because of the type of data being analyzed. Many contractors bid management costs as a fixed cost. But some do not. Still the point of the problem was to demonstrate the power of the absolute deviation method for comparison purposes and I hope that comes through. Regardless of all the tools that have been shown in this text, usually the best answer to a learning curve is determined by seeing what method minimizes the deviation from what was forecast.

Phil's next case will be a bit more tricky. Here's the data set:

Lot	Total	Units	Constant Unit Cost
Lot 1	$30,000,000	100	$300,000
Lot 2	$45,097,800	300	$150,326
Lot 3	$39,623,100	300	$132,077

Let's see what our tools can do with this data set:

The nonlinear regression model shows the following:

Year	Lot	Units	Cost	Unit Cost
	1996	1	100 $30,000,000	$300,000
	1997	2	300 $45,097,800	$150,326
	1999	3	300 $36,923,100	$123,077

Deesc	Dollars	Total
100.00%	$300,000	$30,000,000

100.00% $150,326 $45,097,800
100.00% $123,077 $36,923,100

Lot	Midpt X	Y	Log X	Log Y
1	32.35	$300,000	1.510	5.477
2	228.70	$150,326	2.359	5.177
3	541.36	$123,077	2.733	5.090
			6.603	15.744

Log X Log Y	(log x)^2
8.270	2.280
12.214	5.566
13.914	7.472
34.398	15.318

R Sq
99.79%

A = $907,936
LC =   79.96%

I have a very good fit and a potential solution.

The average cost extrapolation method provides the following:

	Units	Total Cost
Lot 1	100	30,000,000
Lot 2	600	84,720,900

Learning Curve: 80.62%
A: $864,843

I use these two method to determine my absolute deviation from my data set:

Mean Average Deviation Calculation for Nonlinear regression

Lot	Actual	Predicted	Units	Total Actual
1	$300,000	$295,808	100	$30,000,000
2	$150,326	$157,405	300	$45,097,800
3	$123,077	$119,208	300	$36,923,100
				$112,020,900

Total Predicted	Total Error
$29,580,771	$419,229
$47,221,456	$2,123,656
$35,762,337	$1,160,763
	$3,703,648
AD	3.31%

Average Cost Extrapolation Method

LC: 80.62%
b: -0.3108
A: $864,843

Lot	Units	midpt	Unit Actual	Unit Predicted
1	100	32.49	$300,000	$293,157
2	300	228.82	$150,326	$159,819
3	300	541.41	$132,077	$122,288

Total Actual	Total Predicted	Total Error
$30,000,000	$29,315,689	$684,311
$45,097,800	$47,945,810	$2,848,010
$39,623,1	$36,686,3	$2,936,70

00	96	4
$114,720, 900		$6,469,02 5
Error		5.64% AD

Both of these values look pretty good. So I say I'll assume the nonlinear regression model is the best fit because it has the least absolute deviation.

I make my projection for the fourth lot of 400 units:

$$Y = A * x_{mid}^b$$

$$Y = \$907,936 * 890.59^{-0.3223} = \$101,712$$

$$\text{Lot Cost} = 400 * \$101,712 = \$40,684,976$$

The actual lot cost was: $48,985,600

My error is ($48,985,600-$40,684,976)/$48,985,600 = 16.9%

This is much to high an error if my learning curve calculations were correct. So I have to determine what is really occurring.

First let's look at the data set (it's already in constant dollars so I don't have to worry about inflation clouding my view).

Constant

Lot	Total	Units	Unit Cost
Lot 1	$30,000,000	100	$300,000
Lot 2	$45,097,800	300	$150,326
Lot 3	$39,623,100	300	$132,077
Lot 4	$48,985,600	400	$122,464

I can definitely see that it is not a fixed cost. The total constant dollars vary substantially.

There are two other possibilities that I can investigate. One is that there is a fixed cost and variable cost in each of the totals and the other is that there is a step down from lot 1 to lot 2. So let's solve for these two possibilities.

I solved the fixed cost and variable cost possibility by using the method outlined earlier in the text. I got the following answers:

LC:	85.32%
b:	-0.229
c:	0.771
A:	$355,458

The absolute deviation for this possibility is:

Absolute Deviation calculation for a fixed cost versus variable cost solution

A:		$355,458	
b:		-0.2291	
Fixed:		$14,405,877	
		85.32% Actual	

Lot	Units	Midt	Cost
1	100	34.00	$30,000,000
2	300	230.22	$45,097,800
3	300	541.98	$39,623,100
			$114,720,900

Predicted Cost	Actual Unit	Variable Unit
$30,253,125	$300,000	$158,472
$45,080,917	$150,326	$102,250
$39,617,647	$132,077	$84,039

Fixed Unit	Predicted Unit	Error
$144,059	$302,531	$253,125
$48,020	$150,270	$16,883
$48,020	$132,059	$5,453
		$275,460
	AD	0.24%

This looks like a pretty good fit and I might be very happy to accept this. But I also can evaluate a step down to see if it is a good fit also. Remember a step down is a lowering of the value from the original A to a new A while keeping the learning curve the same. While I do not know what the learning curve really is I can solve for several step downs, check the resultant learning curve and AD and see if I can get close to a decent answer for the last three lots.

Let's try a 25% step down.

The A value for various learning curves are:

Lot		midpt	Unit Cost			
	1	32.27	$300,000			
					25% New	
LC		b:	A		Step Down	A
95.00%		-0.074	$387,948		$96,987	$290,961
90.00%		-0.152	$508,702		$127,175	$381,526
85.00%		-0.23447	$677,458		$169,364	$508,093
80.00%		-0.32193	$918,007		$229,502	$688,505
75.00%		-0.41504	$1,268,614		$317,153	$951,460
70.00%		-0.51457	$1,792,705		$448,176	$1,344,528

For convenience I used a common midpoint so I introduced some error but it is sufficient for my purposes.

Now I need to run my AD and see if any of these values are close. For instance here is the run for a 95% learning curve with a 25% step down.

95.00% with a 25% step down

Lot	Midpt	Units	Projected	Actual
2	228.62	300	$194,652	$150,326
3	541.33	300	$182,624	$132,077
4	890.59	400	$176,018	$122,464
	Total			

Actual	Error
$45,097,800	$13,297,776
$39,623,100	$15,163,959
$48,985,600	$21,421,529
$133,706,50 0	$49,883,265
	37.31% AD

After repeated iterations I finally get the following answer:

90.00% with a 32% step down

Lot		Midpt	Units	Projected	Actual	Total Projected
	2	228.62	300	$150,491	$150,326	$45,147,427
	3	541.33	300	$132,011	$132,077	$39,603,263
	4	890.59	400	$122,390	$122,464	$48,955,811

Total

Actual	Error
$45,097,800	$49,627
$39,623,100	$19,837
$48,985,600	$29,789
$133,706,500	$99,253
	0.07% AD

To get the values as precisely as possible I change the midpoints to a 90% midpoint and I have my solution:

90.00% with a 32% step down

Lot	Midpt	Units	Projected	Actual	Total Projected
2	231.54	300	$150,201	$150,326	$45,060,415
3	542.51	300	$131,967	$132,077	$39,590,157
4	891.87	400	$122,363	$122,464	$48,945,124

Total

Actual	Error
$45,097,800	$37,385
$39,623,100	$32,943

$48,985,600 $40,476
$133,706,500 $110,803
          0.08% AD

Will other values work?

No value works for 95% (do a goal seek and test this out).

For 85% the best AD is:

85.00% with a15% step down

Lot	Midpt	Units	Projected	Actual	Total Projected
2	231.54	300	$161,425.6	$150,326	$48,427,687
3	542.51	300	$132,212.2	$132,077	$39,663,658
4	891.87	400	$117,666.3	$122,464	$47,066,500

Total

Actual	Error
$45,097,800	$3,329,887
$39,623,100	$40,558
$48,985,600	$1,919,100
$133,706,500	$5,289,545
	3.96% AD

Which is close.

For 80% the best AD is:

80.00% with a -10% step down

Lot	Midpt	Units	Projected	Actual	Total Projected

2	231.54	300	$175,554	$150,326	$52,666,297
3	542.51	300	$133,465	$132,077	$40,039,593
4	891.87	400	$113,728	$122,464	$45,491,066

Total

Actual	Error
$45,097,800	$7,568,497
$39,623,100	$416,493
$48,985,600	$3,494,534
$133,706,500	$11,479,524
	8.59% AD

Since this is a negative step down I need to go no further. The best fit for this particular data set is around a 90% learning curve with a step down of about 32%.

I have two answers that are about equivalent in accuracy (fixed cost plus variable cost method and the step down method). I would probably make a projection for both and see if there was any significant difference for the next lot. If there was I would have a decision to make as to which to report to management.

If you have figured all of this out, reward yourself with a gimlet like Marlowe used to do.

How to use learning curves to determine problems in the system

Learning curves can be used to determine areas in which things are going very right and areas where things are going very wrong. For instance:

Department A:

Month	Projected Hours	Actual	Predicted	Made
1	188	190	60	60
2	152	300	60	40

Right away I know something is wrong in Department A. They are taking much too much time to complete the 60 units allocated to them. Now only are they behind schedule on the number of units but they are expending about twice as many hours as they should in making the parts.

Firms always have methods of determining how many units have been produced and how many were supposed to be produced, however, many fail to consider how many hours were predicted versus how many hours were expended. For instance, if my example had shown that 40 units had been made but that only 120 hours had been expended there may not be a problem with the method but rather that the supervisor had not put enough people in the department to accomplish the work. If you do a comparison based on projected hours and actual hours it leaps right out at you.

Learning curves can be used to give you a rough estimate of what your efficiency is for a given time period. Every firm defines efficiency in their own way so I'll define it simply (planned hours/actual hours) for this example.

Efficiency for month 1 becomes: 188/190 = 98.9%
Efficiency for month 2 becomes: 152/300 = 50.7%

Chapter 16 The Effect of Restarting Learning?

Let's assume the following has happened:

My vendor has produced 1,000 widgets. The vendor has consistently followed a 90% learning curve in the past but we do not have a contractually agreed to learning curve. I have the following:

First lot: 1,000 widgets, average cost: $300 actual

Second lot: 1,000 widgets, average cost $150 bid

Now I had performed an estimate prior to his bid of $150 per widget and I had come up with the following:

$A = Y_{avg\ cost} / Y_{mid}{}^b = \$300 / 340.60^{-0.152} = \$728$

The unit estimate for the second lot of 1,000 widgets:

$\$728 * 1{,}467.65^{-0.152} = \$240$

I am off by $240-$150 = $90 per unit or $90 K and my boss is not happy.

Why?

Well there can be a couple of reasons. One the contractor is actually operating to a much steeper learning curve. Having used this vendor often in the past I am reluctant to believe this.

Another reason could be that the contractor has gotten wind that we are looking into another vendor at a lesser price and he has voluntarily agreed to reduce his price to keep the work. This is more likely.

Another reason could be that he has gotten a much larger order from another source and he will be producing many more widgets and as a result we are going down the learning curve much quicker.

Still a fourth reason could be that he has changed his method.

Let's assume that I have investigated and found that the vendor has in fact changed his method of operation. Now I have a problem. How do I estimate the next lot (assume it is 1,000 units)? Do I assume a new learning curve and project from the current lot or do I assume that the new method is just an extension of the current lot and therefore determine a new A value based on the $150 for the second lot and use that to project the next lot. It all depends upon the details, but I will show you how you can restart learning which is the reason I came up with this long winded problem to start with.

Let's say we are going to assume that the contractor will continue to follow a 90% learning curve and that his previous 1,000 units will count. Here's how I get my A and determine my projection.

A will be based on the midpoint of 2,000 units not 1,000 units, when I am counting all the units in the learning equation:

Here's the solution: $A = Y_{avg\ cost}/Y_{mid}{}^b =$ $\$150/\ 1{,}467.6^{-0.152} = \$455$

My prediction for the next lot unit cost (another 1,000 units):

$\$455 * 2{,}481.12^{-0.152} = \$139/unit$

Now let's compare this with an example where I assume a new learning curve is appropriate:

$$A = Y_{avg\ cost}/Y_{mid}{}^b = \$150/\ 340.60^{-0.152} = \$364$$

My prediction for the next lot unit cost (another 1,000 units):

$$\$364 * 1,467.4^{-0.152} = \$120\ /\ unit$$

The fact that the unit cost is less for a situation where I have less learning might surprise some people but it's only because they haven't thought through the steps. It's true that I have more learning in the first prediction but I have more learning going both ways (both up and down the curve). It so happens, in this example at least, that the journey up and down the curve provides a smaller cost when I assume the learning has started over. Now this is just a mathematical construct and the vendor will bid what he will bid for the next 1,000 units. But it is something to be aware of when you are considering restarting the learning. Restarting will usually give you a smaller price (not necessarily a better estimate) because you are at the start of the learning curve again where the learning is most rapid.

Chapter 17 How to develop a common learning curve

Suppose we have the following data:

Lot 1: 500 Units
Unit Cost: $8,000
LC: 90% for labor and 93% for material

Also suppose my labor is 40% of the total unit cost and my material is 60% of my first unit cost. The question is can I develop a common learning curve for these two items.

First let's see what we have with our typical learning curve method:

		Units	LC	b:
Labor:	$3,200	500	90.00%	-0.152
Material:	$4,800	500	93.00%	-0.1047
	$8,000			
midpt	A			
171.25	$6,993			
175.82	$8,247			

Now I want to project the next lot (I will ignore inflation) of 800 units

If I use two different learning curves I get the following:

Projection for labor and material seperately for lot 2

	A	Midpt	b:	
Labor	$6,993	864.69	-0.152	$2,502
Material	$8,247	866.18	-0.1047	$4,062
				$6,564

So I have an estimate of $6,564 per unit. My combined learning curve becomes:

LC: 91.86%
A: $15,030

Let's assume we want to project another 600 units (lot 3):

If I use my standard method of breaking labor and material into their separate learning curves I get the following:

Projection for labor and material seperately for lot 3

	A	Midpt	b:	
Labor	$6,993	1,589.62	-0.152	$2,281
Material	$8,247	1,590.06	-0.1047	$3,812
				$6,092

Now suppose we use the joint A and learning curve values we have determined to project the unit cost for the next 600:

	A	Midpt	b:	
Joint	$15,030	1,589.89	-0.12249	$6,093

There is little difference between the two. Once the relationship has been established as to the percentage of the total that belongs to one learning curve and what belongs to another and if that percentage holds then future estimates are just as accurate using a joint midpoint as if we were using two separate midpoints. For instance, assume that we wish an additional 40,000 units for lot 4. The projections for the two methods become:

Projection for labor and material seperately for lot 4

	A	Midpt	b:	
Labor	$6,993	17,132.07	-0.152	$1,589
Material	$8,247	17,359.35	-0.1047	$2,968
				$4,557

And for the joint method:

	A	Midpt	b:	
Joint	$15,030	17,274.10	-0.12249	$4,549

Even with the giant leap in the requirements the costs did not diverge. This little tip is made to order for DTUPC type pricing when the assumption is made that the proportions will remain about the same throughout the life of the project. Once the different percentages are determined, you can easily develop one joint learning curve and use it to forecast your costs.

Now what happens if we have a situation where there are widely varying learning curves. I'll show a 75% labor and a 95% material example to see if we can still use a common learning curve. Here's the set up:

Lot 1: 500 Units
Unit Cost: $8,000
LC: 75% for labor and 95% for material
Also suppose my labor is 40% of the total unit cost and my material is 60% of my first unit cost. The question is can I develop a common learning curve for these two items.

	Units	LC	b:
Labor:	$3,200	500	75.00%	-0.41504
Material:	$4,800	500	95.00%	-0.074
	$8,000			

midpt	A
143.16	$25,113
178.7	$7,045

As you see the A value for labor has increased significantly:

Now let's project for the next 800 units

Projection for labor and material seperately for lot 2

	A	Midpt	b:	
Labor	$25,113	856.34	-0.415	$1,523

Material	$7,045	867.15	-0.074	$4,271
				$5,794

So I have an estimate of $5,794 per unit. My combined learning curve becomes:

LC: 87.38%
A: $21,595

Let's assume we want to project another 600 units (lot 3):

If I use my standard method of breaking labor and material into their separate learning curves I get the following:

Projection for labor and material seperately for lot 3

	A	Midpt	b:	
Labor	$25,113	1,589.62	-0.41504	$1,178
Material	$7,045	1,587.12	-0.074	$4,084
				$5,262

Now we will use the joint A and learning curve values that we determined based on the first two lots:

	A	Midpt	b:	
Joint	$21,595	1,589.21	-0.19462	$5,144

The difference between the two is $5,262-$5,144 = $118 and the error is $118/$5,262 = 2.25%

Our shortcut just doesn't work when there is a wide variance between the learning curves. As a rule of thumb never combine learning curves when there is more than 5% between the two. Of course, the wise engineer will run out a test before using this shortcut to insure that the errors are small and that the estimate is not biased regardless of the percentage.

Chapter 18 How do I handle operations that are tied to a machine?

Assume the following:

Time to produce 1 widget 1 hour
Time to produce 1,000 widgets 1,000 hours

This happens often when the widget time is made up of machine time only.

A machine does not learn. When projecting future savings I may want to use a learning exponent because I believe we will replace the machine with a better one, or improve the operation of the current machine or just because I want the cost to decline for some devious reason.

What does this mean to an estimator? You must find out how much of the time in your estimate is made up of machine process time and whether or not the operator has to stay with the machine during the process. Machine time should be treated like standard hours over the short run. It is appropriate to use a learning curve for improvement for the future but if you do it for short term estimates you will underestimate the cost required to produce the unit.

There will often be tests that are ran on individual units to determine whether those units are ready to pass on to the next stage of production. If these tests are time consuming and require an operator to stay with the test, make sure that a learning curve is not used to estimate the cost of the lot.

Chapter 19 Converting cumalative average learning curves to unit learning curves

There are occasions when you wish to change from one type of learning curve to another. For instance, a contractor might bid mistakenly based on a cumulative average curve and you have to convert that in order to be able to see what the cost would be if the bid had correctly been made using a unit curve. Now the easiest way to do this is to send the blasted bid back and tell the joker to do it right but suppose your boss wants to know right now and you are the one he wants to know from (pardon the English). Well it's not very difficult to transform one into another using the power of a spreadsheet but remember as was shown earlier if you project based on a unit and the contractor projects based on a cumulative average, you will get different projections.

Back to the problem at hand:

Let's set up the example.

A = $15,000
Cum Average LC: 90%

b: -0.152 (log .90/log 2) just like in a unit curve

Lot 1 300
Lot 2 500
Lot 3 600

The unit cost for the first lot would be:

$15,000 * (300^{-.152}) = $6,303

The total cost for the first lot would be:
300*$6,303 = $1,890,900

The unit cost for the first + second lot would be

$15,000 * ((300+500)=800^{-.152}) = $5,430

The cost for the first and second lots together would be:

800*$5,430 = $4,344,000

That means that the cost for the second lot is $4,344,000 - $1,890,900 = $2,453,100

The unit cost for the second lot is $2,453,100/500 = $4,906

The unit cost for the first + second + third lot would be:

$15,000 * ((300+500+600)=1,400^{-0.152}) =$ $4,987

The lot cost for all three lots would be:

$1,400 * \$4,987 = \$6,982,450$

The cost for the third lot is :

$\$6,982,450 - \$4,344,000 = \$2,638,450$

The unit cost for the third lot is:

$\$2,638,450/600 = \$4,397$

Before I go further, I want you to see how easy these numbers were to obtain. No midpoints were required, all I have to know is the last unit, I use the last unit as the projection of my costs. Why most people use unit instead of cum average is beyond me, but they do.

Back to the problem:

I have the following:

Lot	Units	Unit Cost	Lot Cost
1	300	$6,303	$1,890,900
2	800	$5,430	$2,453,100
3	1,400	$4,987	$2,638,450
			$6,982,450

My total unit cost is $6,982,450/1,400 = $4,987

Now my boss wants to know what unit learning cure this represents:

To determine this, I will not know the A value or the LC value used by the contractor (all I have is his bid) and the knowledge that he had used a cum average curve.

To find the A and b values is simple:

$Y_{300} = A(300^b) = \$6,303$
$Y_{800} = A(800^b) = \$5,430$

$\$6,303/\$5,430 = A(300^b)/A(800^b)$

Eliminate the A value and divide the left hand side:

$1.1608 = (300^b)/(800^b)$

Take b outside of the equation:

$1.1608 = (300/800)^b$

Use logarithms to put this into a linear form:

$Log(1.1608) = b*\log(300/800)$

0.0647= b*(-0.426)

b = 0.0647/-0.426 = -0.152

b = Log (lc)/(log 2)

Log LC = -0.152*0.301 = -0.04575

LC = $10^{-.04575}$ = 90%

Again see how easy it is to calculate. No long unit cost equations but rather a direct application of the one basic learning curve equation.

Now I know that he has bid a 90% learning curve (which I will check with the next batch) and I can calculate the A value directly using either the 300 lot or the 800 lot.

Y_{300} = $6,303 = A($300^{-0.152}$)

$6,303 = 0.4202 A

A = $6,303/0.4202= $15,000

Now that I know what the contractor has done I can project any amount I want to and then I will convert the projection to a unit learning curve.

Let's say that all I need to do is to project the next 600 units:

Now I have $15,000 * (1,400^{-0.152}) = $4,987$

Now the A value should not change. Remember it is the cost to perform the first unit and that is the same regardless of how the data is gathered.

I have $A = $15,000$
Unit Cost: $4,987 (which is the same as the midpoint unit cost)
Units 1,400

All I have to do is solve for the standard learning curve equation:

$$Y_{mid1,400} = A\, x^b$$

$$\$4,987 = \$15,000 * x_{mid1,400}^b$$

Remember that the midpoint and b are unique so I really have only one (not two) unknowns.

Now I solve for $x_{mid1,400}^b$

$$\$4,987/\$15,000 = x_{mid1,400}^b = 0.333$$

Now what learning curve at a 1,400 midpoint is equal to 0.333?

Well I can solve this algebraically if I go back to my midpoint calculation section and do a lot of work but instead I choose to call up my midpoint spreadsheet and just click away until I have the answer.

I must have the midpoint and learning curve that will provide me with a 0.3325 proportion between $15,000 and the estimate for the 1,400 units. I completed a table to show you how to iterate to get this number.

Iterate until you have a proportion of 0.333

A	Learning Curve	Midpt	b:
$15,000	75.00%	393.5	-0.41504
$15,000	80.00%	424.53	-0.32193
$15,000	85.00%	451.84	-0.23447
$15,000	90.00%	475.98	-0.152
$15,000	88.33%	468.24	-0.17904

Lot Cost	Percent of A	
$1,256	0.084	
$2,138	0.143	
$3,578	0.239	
$5,876	0.392	Too Far
$4,988	0.333	Stop

So in about 3 minutes on my spreadsheet I have shown that the equivalent unit learning curve for this scenario is 88.33%. I know I have said this before but it is important. If you project the next batch based on a unit curve of 88.33% with an A of $15,000 you will not get the same answer as if you project a 90% cumulative curve with an A of $15,000. But I did answer my boss's question. For an example of converting a unit learning curve to a cumulative average learning curve see the next section.

Chapter 20 How to solve unit learning curves when the first lot information is incomplete

Let's suppose the following:

I know that my company has purchased 3 lots:

The first lot was 100 units
The second lot was 200 units
The third lot was 300 units

I know that I paid the following:

Lot 1: Unknown
Lot 2: $900,630
Lot 3: $1,189,078

Can I determine the learning curve from this partial information?

Yes but there will be some error. Let me demonstrate.

First, you might rightly complain that a company should track the data better than this. I wouldn't disagree, but there are times when you only have partial information available and some jerk er, that is, some nice boss asks you to solve the problem anyway. This trick will work as long as you know the total number of units that have been produced and that information is usually available. Let's see how this works.

Recall what I wrote about midpoints (see Do midpoints matter). The midpoints converge to a common point (and do so fairly quickly) regardless of the learning curve. For instance, here are the midpoints for three unit learning curves for the 600 units under question,

90%	85%	80%
35.39	33.91	32.27
190.3	189.56	188.76
440.65	439.94	439.19

Suppose I choose 85% as the most probable learning curve to use. Now I can set up my equations and solve via logarithms.

$$Y_1 = A \, (x_{mid100})^b$$

$$Y_2 = A \, (x_{mid300})^b$$

$$Y_3 = A \, (x_{mid600})^b$$

$$Y_2 = \$900{,}630/200 = \$4{,}503.15$$

Y_3 = \$1,189,070/300 = \$3,963.57

Let's set up our logs:

Log Y_1 = Log A + b(log x_{mid100})

Log Y_2 = Log A + b(log x_{mid300})

Log Y_3 = Log A + b(log x_{mid600})

Our unknowns are A, b, Y_1 and all the midpoints. But being a brave soul I will use my 85% midpoints and see if it gives me a decent answer.

Log Y_1 = Log A + b(log 33.91)

Log (\$4,503.15) = Log A + b(log 189.56)

Log (\$3,963.57) = Log A + b(log 439.94)

Let's name the equations and solve the logs we do know:

EQ 1: Log Y_1 = Log A + b(1.5302)

EQ 2: 3.6535 = Log A + b(2.2777)

EQ 3: 3.5981 = Log A + b(2.6434)

As you can see we now have 3 equations and three unknowns.

Since our goal is to solve for A and b we really do not need Equation 1 but we will keep it and explore the final results anyway.

First subtract equation 3 from equation 2.

.0554 = -0.3664 b

b = -0.1512

Recall that:

b = Log(LC)/Log(2)

So Log (LC) = b*log 2 = -.1512*.301 = -0.0455

Anti log (LC) = $10^{-0.0455}$ = 90.05%

Now let's solve for A;

Take equation 2 and substitute in the b value:

EQ 2: 3.6535 = Log A + b(2.2777)

3.6535 = Log A +(-0.1512*2.2777)

3.6535 = Log A −0.3444

3.6535+0.3444 = Log A

3.9979 = Log A

A = Antilog(3.9979) = $10^{3.9979}$ = $9,952

Now I started with a 90% learning curve and an A value of $10,000 so I am very close to the actual learning curve.

Let's solve for Y_1

EQ 1: Log Y_1 = Log A + b(1.5302)

Log Y_1 = 3.9979 + (-0.1512*1.5302)

Log Y_1 = 3.9979-0.2314

Log Y_1 = 3.7665

Y_1 = $10^{3.7665}$ = $5,841

So my average unit cost for the first 100 units should be about $5,481.

My actual lot cost for a 90% learning curve and an A of $10,000 for the first 100 units is:

$581,410/100 = $5,814

As you can see my error is: ($5,814-$5,481)/$5,814 = 5.72%

This error can be lessened appreciably if I take the learning curve data I have derived and use it to determine what my Y_1 value should be.

My new midpoint: 35.40 (100 units at a 90.05% learning curve)

My A value is $9,952

My b value is: -0.1512

My projection for the unit cost for the first 100 units becomes:

$Y_1 = \$9,952 * (35.40^{-0.1512}) = \$5,803.65$

Now my error is:

($5,814 - $5,804)/$5,814 = .17%

This method is of course mathematically incorrect (the 85% midpoints introduce error) but still it provides a fairly accurate A and b.

Let's look at mipoints from 95% down to 65% for our example:

LC	95%	90%	85%	80%
100 midpt	36.74	35.39	33.91	32.27
300 midpt	191.01	190.3	189.56	188.76
600 midpt	441.33	440.65	439.94	439.19
75%	70.00%	65.00%		
30.48	28.5	26.34		
187.91	187.01	186.04		
438.38	437.52	436.6		

Now let's look at the base 10 logarithms of these numbers:

Log 10 of LC

LC		95%	90%	85%	80%
Log 100 midpt		1.5651	1.5489	1.5303	1.5088
Log 300 midpt		2.2811	2.2794	2.2777	2.2759
Log 600 midpt		2.6448	2.6441	2.6434	2.6427
75%	70.00%	65.00%			
1.4840	1.4548	1.4206			
2.2739	2.2719	2.2696			
2.6419	2.6410	2.6401			

The 600 midpoint has very little variation (variation in the 4th decimal place). The 300 midpoint has a little more variation (slight variation in the 2nd decimal place). The 100 midpoint has the most variation and that's why it is inadvisable to use this method to determine the average unit cost for the first lot.

I used 85% midpoints and got a very good answer, but I would have gotten a very good answer regardless of which one I used because of the low variation between the midpoints as they converge.

Now can I use the same trick for cumulative average learning curves?

I can't though a better mathematician (and there are very few that are worse) might have a better method than I. Let me show you why.

Again we have the following information:

The first lot was 100 units
The second lot was 200 units
The third lot was 300 units

I know that I paid the following:

Lot 1: Unknown
Lot 2: $900,630
Lot 3: $1,189,078

The form of the Cum Average curve is $Y = Ax^b$

Now this is a cumulative average so I must know the cumulative value of Y. Without knowing what the first lot cost I cannot work out the rest of the variables because I always have more unknowns than I have equations. Here are the first three equations:

$Y_1 = A(100^b)$

$Y_2 = A(300^b)$ where Y_2 includes Y_1

$Y_3 = A(600^b)$ where Y_3 includes Y_1 and Y_2

Now I have 3 equations and 5 unknowns $(Y_1, Y_2, Y_3, A, -b)$

I cannot use logarithms because Y_2 equals a portion of Y_1.

The only way to solve it is to first solve the problem as a unit learning curve and then transform the data into a cumulative average. For instance the cumulative average for the example I worked is:

$Y_1 = A(100^b)$ where Y_1 is the cumulative average.

Now you know the reason I solved for Y_1 in the unit curve example. It allows me to solve for the cumulative average learning curve.

Y_1 = lot cost lot 1/total units in lot 1

Now the lot cost I derived for the first lot was:

Unit cost of $5,804

A = $9,952

Also recall that A is the first unit cost and is the same for both cumulative average and unit learning curves. So the solution to the cum average curve is a simple algebraic manipulation.

$5,804 = $9,952 (100^b)

Solving for b

$5,804/$9,952 = 100^b

0.5832 = 100^b

Log (0.5832) = b * log 100

-0.2342 = 2b

-0.1171 = b

Log (LC)/Log (2) = b = -0.1171

Log (LC) = -0.1171*0.301 = -0.0352

LC = antilog (-0.0352) = $10^{-0.0352}$ = 92.2%

This particular set of data represents a cum average learning curve of 92.2% for the first lot only.

The second lot is solved the same way and it will show a different learning curve.

Y_2 = [($5,804*100) + ($900,630)]/(100 + 200)

$Y_2 = \$1,481,030/300 = \$4,937$

$\$4,937 = \$9,952 * 300^b$

Solving yields:

$b = -0.12289$

$LC = 91.84\%$

Remember that I am using unit learning curve data to solve points on a curve for the cumulative average. The learning variables for the cumulative average will always change as I go from lot to lot since the analysis data is based on a unit curve.

Chapter 21 Developing a learning curve for set up times or other sporadic activities

There are many jobs that are completed only sporadically (set up for instance) and the learning to is sporadic. Depending on how often the activity occurs, people will forget some of the tricks they have learned, and they will not be as comfortable working on the job until they have spent a certain amount of time working. For the most part, the organizations I have worked for have just ignored this and used the same learning curve for set up time (or any other sporadic event) as was used for all the other work. Whether to develop an individual learning curve for this sporadic work depends on a lot of factors (for instance, how big a part of the total job is the set up time?) and you can spend more time trying to obtain data to properly set the learning curve than the accuracy warrants.

If you want to account for this learning effect in your forecasts but do not want to spend a large amount of time in gathering information I recommend the following:

1. Find out how many units are made on average between the occurrences of the activity.

2. Time the activity at least once and determine how many units have been made to the point you have timed the activity.

3. Develop you're A value based on total units and project down a learning curve about 2-3% less steep than the learning curve you use for standard work.

As always, here's an example:

1. The activity occurs every 50 units.

2. I timed it and found out it took 2 hours to do the activity. There have been 150 units made at the point where I timed the activity.

3. The A value is: $Y = Ax^b$. My standard learning curve is a 90% unit learning curve. I will assume a 92.5% learning curve for this sporadic activity.

2 hours/50 units =0.04 hr/unit = $A(x_{mid}^{-01125})$ and $x_{mid} = 53.51$ (150 units at 92.5%)

$A = 0.04/(53.51^{-0.1125}) = 0.0625$ Hrs

The time projected for the first 50 units would be:

$Y_{50} = 0.0625(18.58^{-0.1125}) * 50 = 2.25$ hours

Of course you can set up a data log and time the activity repeatedly and make a much better estimate and you should if it is critical for some reason. But this method allows you to obtain a decent estimate of the time required without doing a lot of effort.

Chapter 22 Cost to Complete and DTUPC Advice

The purpose of this section is to show the reader how to calculate a cost to complete estimate for a program. This type of estimate is common in defense industry work and it is a rather simple estimate to prepare. Normally a cost to complete estimate is a full up (all costs) estimate of what the firm will sell the item for over the life of the contract. There is nothing special about the learning curves (it is just a straight forward application of either a unit learning curve or a cumulative average learning curve) where the learning curves are (usually) specified by the Government or through joint agreement. The estimate can be in constant dollars or current (escalated) dollars dependent upon the needs at the moment (you can obtain definitions for how the Government requires these dollars be calculated by referring to the Cost Analysis Manual for whichever organization is applicable).

A simplified but typical cost to complete spreadsheet might look this way:

Units	600	700	800

	Actual	Projected	Projected
	Lot 1	Lot 2	Lot 3
Item 1	$200,000	$185,591	$193,671
Item 2	$300,000	$243,110	$241,415
Item 3	$350,000	$324,784	$338,924
Assy Labor	$150,000	$104,160	$98,110
Mgt	$100,000	$69,440	$65,406

800
Projected

Lot 4	Total
$182,484	$761,746
$220,223	$1,004,748
$319,346	$1,333,055
$86,469	$438,738
$57,646	$292,492
	$3,830,779

The cost to complete for this example is $3,830,779.

The learning curves and A values are as follows:

LC	b:	midpt	Units
90.00%	-0.152	205.14	600
85.00%	-0.23447	195.09	600
90.00%	-0.152	205.14	600
80.00%	-0.32193	183.8	600
80.00%	-0.32193	183.8	600

Unit Cost	A
$333.33	$748.72
$500.00	$1,721.67
$583.33	$1,310.25
$250.00	$1,339.36
$166.67	$892.90

I calculated this example in a spreadsheet in about 5 minutes. If I wanted to put this example in current year dollars (includes inflation) I would just multiply each of the constant year dollars by the inflation rate for that year.

Notice that the cost to complete shows the planned buy for the lot, the total cost for the lot and specifies which item those values are associated with. Often unit costs are used instead of total costs.

As I pointed out in the section called "Do Midpoints Matter" that a common midpoint can be used for an estimate like this if you are in a time crunch. I have found that calculating the midpoints is by far the most time consuming part of doing a big estimate.

I usually set up a spreadsheet that track cost by each WBS (work breakdown structure) number. This makes it very easy to change some characteristic (except midpoints) and rerun the analysis. I do not attach a midpoint calculator for each WBS however because it slows the speed that I can calculate the spreadsheet tremendously and also it requires more computer size. People do not like to receive email attachments of several megabytes.

A data base can be used to do this type of analysis also, but I have found that there are so many changes that have to be tracked that it is easier to use a spreadsheet for that purpose.

Here is a typical spreadsheet tab for the above example:

Units		600	700	800	800
	Actual	Projected	Projected	Projected	
	Lot 1	Lot 2	Lot 3	Lot 4	
Item 1	$200,000	$185,591	$193,671	$182,484	

Total	LC	b:		
$761,746	90.00%		-0.152	
		Unit		
midpt	Units	Cost	A	
205.14	600	$333.33	$748.72	

Now each of these items are keyed to another spreadsheet tab. I will have a schedule spreadsheet, a learning curve spreadsheet, an A spreadsheet etc. This allows me to change only one item (say a learning curve value) in one place and that item will be instantly recalculated across all the spreadsheets.

A similar type of estimate is called a Design to Unit Production Cost (DTUPC). In this type of estimate, there are specific rules that state what is to be calculated (for instance NRE and tooling are often excluded) and there are goals for each WBS item. DTUPC is typically tracked to a base year dollar.

Assume that the DTUPC goal for Item 1 is $300/unit in FY 2001 dollars.

Assume that lot 1 is in 2001 dollars. How am I doing on making my goal?

Well my unit cost for item 1 is $200,000/600 units = $333.33 per unit so that means I'm not meeting my goal right? Not necessarily, usually the goal is an AUC (Average Unit Cost) and my projections across all 4 lots show that my AUC for item 1 should be:

$761,746/(600+700+800+800) units = $761,746/2,900 units = $262.66 /unit

So even though I am above my goal currently I am on well on track (assuming my learning curve projections are correct) to meet the goal for this particular item.

Chapter 23 Why Learning Curves Can Drive You Crazy

Here is an example of why an engineer is sometimes thought to be an idiot (this time unjustly but not always of course).

Assume the following:

Lot 1: $2,085,949 for 300 units
Lot 2: $1,426,895 for 300 units

I solve this using my handy dandy logarithms that I showed in the lot to lot chapter and get the following:

LC: 84.7%
b: -0.2481
A: $21,522

I am happy, the boss is happy and the vendor is very happy (if he gets this information). You see he thinks his learning curve is:

A: $12,000
Lot 1 LC: 92.2%
b: -0.1172
Lot 2 LC: 90%
b: -0.1520

Don't believe me. Well you can use midpoints to show some values to support your position either way.

The total cost for both lots is $3,512,844 The midpoints for the 84.7% learning curve are:

Lot 1: 97.74
Lot 2: 439.82

The unit cost for lot 1 is: $21,522 * $97.74^{-0.2481}$=$6,905
The unit cost for lot 2 is: $21,522 * 439.82^{-0.2481}$ = $4,754

The total cost for the two lots are:

Lot 1: $6,905*300 = $2,071,500
Lot 2: $4,754*300 = $1,462,200

The total estimate cost is $3,497,700 (compared with an actual cost of $3,512,844) an error of only 0.43%

Now let's look at the other situation. I have two different learning curves operating and I know my first unit cost is $12,000

My unit cost for lot 1 is: $12,000 * 105.4^{-0.1172}$ = $6,952

My unit cost for lot 2 is: $12,000 * $440.65^{-0.152}$ = $4,756

Lot 1 cost is: $6,952*300 = $2,085,600

Lot 2 cost is: $4,796*300 = $1,438,800

The total estimate cost is $3,524,400. The error is ($3,512,844 - $3,524,400)/$3,512,844 = 0.33%

What is happening? Well the learning curve standard format assumes one learning exponent (not multiples). You cannot constantly change learning and expect to obtain a correct answer. The vendor probably did not even have an A value, but was rather adding up his costs for what he was doing, putting his overhead, profit etc. on those costs and bidding the results. To effectively use unit and cum average learning curves I must have an A value that is tied to the production cost.

Once you get a third lot in you can go to the multilot analysis techniques elsewhere in this text. But again, the assumption is that you have both an A value and a continuous learning curve operating.

If either of these assumptions is incorrect then the learning curve calculations will not work. For instance, the vendor may come in with a higher cost for lot 3 than for either lot 1 or 2. If you use a varying learning curve per lot, you might convince yourself that learning is still occurring. Well it isn't. There is absolutely no way for the unit cost to go up (ignoring inflation as always) according to either the cumulative or unit learning curve theories.

Here's a table with a negative learning for the third lot:

A: $12,000

Lot	Midpt	LC	b:
1	105.4	92.20%	-0.11716
2	440.65	90.00%	-0.152
3	745.31	98.00%	-0.0291

Unit Cost	Lot Cost
$6,953	$2,085,949
$4,756	$1,426,895
$9,896	$2,968,823

Based on his estimating methodology (an incorrect one) he can bid a higher unit cost for the 3 bid and still expect to claim some amount of learning. He can't. If the unit cost goes up (after inflation has been removed of course) then there is no learning.

The purpose of this example is to show the reader that there are idiots that calculate learning curves just as there are idiots that do most everything else. Be aware that a true learning curve requires continuous unit price decreases and that there is no such thing as lot to lot increases of unit cost for either of the standard learning curve models.

Chapter 24 Simplistic Trend Analysis

There are times when the boss will come by your desk and say what progress are we making on the so and so contract? Usually these means (in boss speak) work up an analysis on the so and so contract and let's see if we are meeting our objectives.

Let's assume the following:

I am going to order 10 lots of 500 units each from vendor A. Vendor A has agreed to attempt to meet an 85% unit learning curve over the course of the 10 buys but does not guarantee the learning curve. As always I am ignoring inflation in this analysis but you shouldn't when you do the real thing. Now I have the following:

3 lots have been purchased. They came in at:

Lot 1 : $3,000,500
Lot 2: $2,505,650
Lot 3: $2,105,660

Now is the vendor making progress?

There are a number of ways to attack this problem. The way I choose is to determine the A value and the b value for an idealized 85% unit learning curve and then develop a trending format based on the lot to lot cost to see if the vendor is making any progress.

First I have to determine my A value. To do justice to the vendor I will take his data and determine the learning curve and A value for the first two lots. Please see the lot to lot learning curve chapter to see how to do this analysis.

LC: 91.75%
A: $11,370

Now the A value should be the same (first unit cost) regardless of the learning curve. Now I can take the A value and determine what the contractor has to bid to meet an 85% unit learning curve.

A:	$11,370
LC:	85.00%
b:	-0.234465

Lot	Midpt	Unit Cost
1	163	$3,444
2	733	$2,421
3	1,240	$2,140
4	1,743	$1,976
5	2,245	$1,862

6	2,746	$1,776
7	3,247	$1,708
8	3,747	$1,651
9	4,247	$1,604
10	4,748	$1,562

Units	Total Cost
500	$1,722,213
500	$1,210,534
500	$1,070,091
500	$987,988
500	$931,102
500	$888,140
500	$853,934
500	$825,703
500	$801,789
500	$781,126
	$10,072,620

Next I set up a table showing me the values I currently have:

Lot	Cost	85% Bid Cost	Delta
1	$1,722,213	$3,000,500	$1,278,287
2	$1,210,534	$2,505,650	$1,295,116
3	$1,070,091	$2,105,660	$1,035,569
		$7,611,810	$3,608,972

The contractor has already been paid $7,611,810 out of a maximum of $10,072,620. In other words he has not met an 85% learning curve to date and he will have to sell the remaining 3,500 units to us at a cost of ($10,072,620-$7,611,810)/3500 = $703/unit in order to meet an 85% learning curve for the 10 lots.

There might be some grumbling from other learning curve users that I am misusing the A value. It is after all just a mathematical construct that I have calculated using a best fit curve. Well that is true and it is best to have an agreed to A value between you and the vendor after the first lot is delivered. That rarely happens in real life however and the true A value (if there is such a beast) should be fairly close to the value determined from the first two lots.

Chapter 25 Developing a group learning curve

How would you handle the following situation?

I have a work cell that includes 4 people. The work cell produces a new widget that is similar to a gadget we produced before.

My four people are:

Joe-An old hand that has been working for years on items like the widget.
Jan- A new hire.
Julie-A rehire that used to be very good at producing gadgets.
John-A guy with about 3 months of experience in producing gadgets.

Now my job will be to estimate how many hours will be required to produce 1,000 widgets.

My gadget information is as follows:

Lot	Units produced	Time to produce
1	20,000	32,000 hours
2	30,000	34,000 hours

As always the boss wants the estimate before the end of the day and he wants me to base the estimate on unit learning curves because that's what we will be committed to during the contract.

The first thing I do is to determine what the learning curve and a value was for the gadgets. I'll use the average cost method that I have shown several times previously.

The average cost method yields the following:

A: 10.11 hours
LC: 86.46%
b: -0.2099

Now the first thing I need to do is to do an estimate of my A for the widget. It is true it is similar to the gadget but I judge that the widget is slightly more complex. I'll assume that it will be 20% more time consuming and therefore I'll use an A value of 10.11*1.20 = 12.13 hours. I can easily get a value for my estimate by using the 86.46% learning curve and the new A value and an 86.46% midpoint $(12.13 * 328.72^{-0.2099}) \times 1{,}000 = 3{,}594$ hours. But is this the best estimate?

Let's assume that I know that only experienced operators worked on the gadget production line. We had in fact come down the learning curve much faster than we had anticipated. Given this information I might want to do a more detailed evaluation of the workforce for the widgets.

Now I have a problem that requires developing a learning curve for a group of 4 workers. What I have is a mix of experienced and non-experienced people. How do I handle this?

Well this is the type of question that estimators face often and here's how I suggest you handle the problem.

First off you should estimate the potential learning curve for each of the four workers:

Here's my estimate:

Joe- The old hand (85%)
Jan- A new hire (95%)
Julie-A rehire (90%)
John- Slight experience (92%)

This is all of course subjective and it will be wrong. Whether it will provide a better estimate than using the gadget learning curve will only be determined after the fact. But I would rather estimate high, than low. If you have been around long, you will often see labor hours being negotiated away without any consultation with the estimator and yet when this happens, management seldom remembers when it comes time to place accountability for a project running overbudget.

So how do I develop a joint learning curve for this group of four?

Let's see what the hours would be for each of the workers if they performed the work by themselves and determine an average.

Joint Unit Learning Curve Example

Operator	LC	midpt	b:	A:
Joe	85.00%	323.5	-0.2345	12.13
Jan	95.00%	355.8	-0.0740	12.13
Julie	90.00%	340.6	-0.1520	12.13
John	92.00%	346.9	-0.1203	12.13

Units	Unit Estimate	Lot Estimate
1,000	3.13	3,128.77
1,000	7.85	7,853.59
1,000	5.00	4,999.79
1,000	6.00	6,001.92
Total		21,984
Average		5,496

So if my estimate of the relative abilities of the operators are correct then I should bid 5,496 hours for the first lot not 3,594.

Once you start gathering data, these types of problems become less significant. You will no longer have to handicap people, but rather you will be able to take the data for the group as a whole and develop a group learning curve that covers the whole group. This is exactly how the first learning curve papers were developed. The authors took the group results and developed the now classic $Y = Ax^b$ equation.

You can also set up an expected value analysis to determine an estimate. Let's look at how you would do this.

Again, I have to estimate the number of hours for 1,000 widgets. I have the gadget information. From it I develop the same learning curve characteristics:

A: 10.11 hours
LC: 86.46%
b: -0.2099

Now I know that I have only one data point and that projections based on only one data point are often highly inaccurate. To help me better forecast the hours required for the widget I have talked to several people and gotten the following information:

The widget will be 20% more difficult than the gadget to produce.
The widget will be 10% more difficult than the gadget to produce.
The widget will be about the same difficulty as the gadget.
The widget will be about 10% less difficult than the gadget to produce.

I'm totally at a loss of course, I've talked to 4 experienced people and I've gotten 4 different opinions (not an unheard of experience). So I will use the 4 different opinions to develop my learning curve characteristics. What I have is a very loosely defined expected value application (the probability of something occurring multiplied by the value of the occurrence). Here's a spreadsheet I set up to calculate the A value and the learning curve.

Expected Value Unit Learning Curve Example

Opinion	Estimated Percent Above or Below	1 + Col 2	Gadget A	Widget A
1	20.00%	120.00%	10.11	12.13
2	10.00%	110.00%	10.11	11.12

3	0.00%	100.00%	10.11	10.11
4	-10.00%	90.00%	10.11	9.10

	Weighted
Probability	Average
25.00%	3.03
25.00%	2.78
25.00%	2.53
25.00%	2.27 Expected
100.00%	10.62 Value

I have chosen to give each estimate a probability of 25%. I could have chosen 50% for one and split the other 50% between the others or any other method that does not result in a probability summation of greater than 1. I can follow the same process for the learning curve and I can develop an estimate from whatever those values happen to be. The point of this chapter is that you should obtain as much information as possible and to use that information to better your estimate. There are many other techniques available to do this but these can help you when you are under the gun to get out an estimate in a hurry.

Now here is an interesting example that shows a situation that a lot of experienced estimators miss. Here's the setup:

I have a work cell with 5 people

I have a historic learning curve of 90% (unit)

I want to bid a lot of 300

My First unit time is 20 hours

How would you handle this?

Most estimators would do the following:

$$Y_{300} = A*(x_{midpt}{}^b)$$

In this case A = 20 hours
X_{midpt} = 103.42
b = log(.90)/log(2) = -0.152

My projection would be: Y_{300} = 20*(103.42$^{-0.152}$) = 9.88 hours/unit
Total lot time = 300*9.88 = 2,964 hours

Now what is wrong with this?

If the work cell is set up where each individual does one widget then you have overestimated the learning. You do not have learning for 300 units because that is achieved only if all the units go through the same process one after another. In this case you have 5 people in the work cell doing the very same work. Let's assume each is operating to a 90% learning curve. Then the total learning for each employee (and therefore for the lot) is:

300/5 = 60 units

$Y_{60} = A*x_{midpt}{}^b$

In this case:

A = 20 hours
X_{midpt} = 21.69 (midpoint for 60 at 90%
b = log(.90)/log(2) = -0.152

The total lot time becomes:

20 * $21.69^{-0.152}$ x 300 = 3,759 hours

By not realizing that the learning attaches to the unit (and there are only 60 equivalent units since 5 people are doing the very same thing) I have underestimated by 3,579 – 2,964 = 615 hours.

The only time this is a significant problem is for the first or maybe the second lot (assuming a few hundred per lot) because once you have the data collected you will know the group learning curve and you can use that to project into the future. In this instance, the group learning curve was:
$Y_{300}=A(x_{mid}{}^b)$

Y_{300} = 3,579/300 = 11.93 hours

$A = 20$

Now I have to solve for x_{mid}^b and from that determine my learning curve. As was shown earlier in this text x_{mid}^b can be determined by iteration on a computer. For this example:

$11.93/20 = x_{mid}^b$

$0.5965 = x_{mid}^b$

$x_{mid}^b = 105.75^{-0.1190} = 0.5965$

Unit Learning Curve = 92.60%

Chapter 26 Learning curve percentage reductions

Following is a table that shows the cumulative average reductions for learning curves for various percentages and various units. Please note that the steeper the learning curve the higher the reduction. I wanted to show this because it will help the reader when he has to make a gut feel call (as often happens). For instance, in a meeting management will often show you two numbers and ask what's your gut feel on this learning curve?

Cumulative Average learning values

Learning Curve	b	Unit 1	5	10	25	50
99.00%	-0.01449957	1	97.69%	96.72%	95.44%	94.49%
95.00%	-0.074000581	1	88.77%	84.33%	78.80%	74.86%
90.00%	-0.152003093	1	78.30%	70.47%	61.31%	55.18%
85.00%	-0.234465254	1	68.57%	58.28%	47.01%	39.96%
80.00%	-0.321928095	1	59.56%	47.65%	35.48%	28.38%
75.00%	-0.415037499	1	51.27%	38.46%	26.29%	19.72%
70.00%	-0.514573173	1	43.68%	30.58%	19.08%	13.36%
65.00%	-0.621488377	1	36.78%	23.91%	13.53%	8.79%
60.00%	-0.736965594	1	30.54%	18.32%	9.33%	5.60%

100	1,000	2,000	5,000	10,000
93.54%	90.47%	89.56%	88.38%	87.50%
71.12%	59.98%	56.98%	53.24%	50.58%
49.66%	34.99%	31.49%	27.40%	24.66%
33.97%	19.80%	16.83%	13.57%	11.54%
22.71%	10.82%	8.66%	6.44%	5.16%

14.79%	5.69%	4.27%	2.92%	2.19%
9.35%	2.86%	2.00%	1.25%	0.87%
5.72%	1.37%	0.89%	0.50%	0.33%
3.36%	0.62%	0.37%	0.19%	0.11%

Well your best guess answer is often incorrect (at least mine is) but I can rule out anything below a 60% cumulative average learning curve because I've never seen a process consistently operate at below a 60% cumulative average learning curve.

Here is the same table for a unit learning curve:

Unit Curve learning values

Learning Curve	b	Unit 1	5	10	25	50
99.00%	-0.0145	1	98.65%	97.85%	96.70%	95.80%
95.00%	-0.0740	1	93.33%	89.60%	84.40%	80.46%
90.00%	-0.1520	1	87.00%	80.06%	70.90%	64.31%
85.00%	-0.2345	1	80.97%	71.35%	59.28%	51.07%
80.00%	-0.3219	1	75.24%	63.44%	49.35%	40.30%
75.00%	-0.4150	1	69.90%	56.25%	40.90%	31.62%
70.00%	-0.5146	1	64.85%	49.77%	33.80%	24.71%
65.00%	-0.6215	1	60.08%	43.98%	27.88%	19.25%
60.00%	-0.7370	1	55.74%	38.91%	22.97%	14.99%

100	1,000	2,000	5,000	10,000
94.87%	91.79%	90.88%	89.68%	88.79%
76.59%	64.75%	61.52%	57.49%	54.62%
58.15%	41.22%	37.11%	32.30%	29.07%
43.77%	25.79%	21.95%	17.72%	15.07%
32.68%	15.87%	12.72%	9.49%	7.59%
24.21%	9.61%	7.24%	4.96%	3.73%
17.84%	5.74%	4.05%	2.54%	1.79%
13.09%	3.41%	2.24%	1.29%	0.84%

9.62%   2.02%   1.25%   0.65%   0.40%

As you can easily see the cumulative average goes down faster than the unit for a given lot. There is nothing wrong with this, they are both just conventions and having different assumptions they will of course have different values. Another thing to notice in the unit curve is the fact that the percentages for the 100-1,000 lots for the 90% and 85% curves go from 58.15% to 25.79%. Many management people will make a statement that the first lot learning is usually about 1/3 of the first unit cost. They know this from past experience. Usually their experience is for small lots and for learning curves between 85% and 90%. It is a very good rule of thumb. If the first lot does not yield a 67% savings then you should investigate further.

Chapter 27 How Long Until It's Done?

You can determine when a batch will be completed by using learning curves.
Here's how:

Assume that I want to produce a batch of 300 widgets

Being very smart I already know the following:

A: 20 Hours
LC: 90% so b =-0.152

Now I want to know how long it will take me to build the 300 widgets and I will show you how to do it using both unit and cumulative average learning curves..

Unit average solution:

Let's first find the midpoint for a lot of 300 units and a learning curve of 90%. This value is 103.42

$Y_{avg} = 20*103.42^{-0.152} = 9.88$ hours

Total lot hours = 300 units * 9.88 hours/unit = 2,964 hours

Now this is just the production hours. I must know the number of workers that will be used to produce the units and the efficiency that I use for those workers and the number of hours per day that they will work.

Let's assume I have 5 people and that my efficiency is 85% and that each person works 8 hours per day.

8*.85 = 6.8 hours of productive work per person.

5 * 6.8 = 34 hours of productive work per day.

2,964/34 = 87.2 production days to produce the batch of 300 units.

Cumulative average solution

I am going to assume the same learning curve for the cumulative average example (but of course it would be different if I was working from actual data) for ease of calculation.

A = 20 hours
LC: 90%, b = -0.152
$Y_{avg} = 20*300^{-0.152} = 8.40$ hours/unit

Total hours for the lot = 300*8.40 = 2,520 hours

Total production days required: 2,520 hours/34 hours/prod day = 74.1 days

Chapter 28 Cost of Quality

Learning curves can be used to determine (roughly) the labor cost of poor quality. Here's how:

Let's assume the following:

A= 10,000
LC: 90%
b: -0.152

Lot	Units
1	300
2	400
3	500
4	600

I am going to assume that all of these lots fall in one year so that I can ignore inflation. Now I can develop a profile for what I should be achieving with my learning curve:

Lot	Units	Midpt	Unit Hours
1	300	103.42	4,941
2	400	484.55	3,907
3	500	937.67	3,534
4	600	1,488.88	3,294

Let's say that my actual results are as follows:

Lot	Units	Unit Hours
1	300	7,200
2	400	6,900
3	500	4,300
4	600	3,800

Now I can determine my cost of quality (or actual efficiency) depending upon how I interpret the data. If I think all of my problems are caused due to poor quality (and not poor methods) then it is a simple subtraction of the two to determine what poor quality has cost the firm.

Cost of Poor Quality

Lot	Units	Unit Forecast	Unit Actual	Total Forecast
1	300	4,941	7,200	1,482,160
2	400	3,907	6,900	1,562,717
3	500	3,534	4,300	1,766,889
4	600	3,294	3,800	1,976,363
	1,800			6,788,129

Total Actual	Cost of Quality
2,160,000	677,840
2,760,000	1,197,283
2,150,000	383,111
2,280,000	303,637
9,350,000	2,561,871

The quality problem has cost me
2,561,871/1,800 = 1,423 hours/unit.

This assumes that all of my learning
problems are associated with quality
problems and that rarely occurs but still it
allows you to come up with a
measurement tool that can aid you not
only in targeting learning problems but
also in targeting quality problems

Chapter 29 Solving Learning Curves without knowing the first lot

Earlier in this text I solved a problem where I did not know the prior cost of units but I did know the number of prior units. Now I am going to solve a problem where I do not know either the cost of prior units or the number of prior units.

This is a very oddball type of problem and I doubt that you will ever have to solve one similar but I once solved this (not for fun but because I was challenged to by a fellow engineer) and so I am bound and determined to show the process.

Here's the setup:

I have the following information:

Lot prior: unknown
Lot 2 : 200 units, unit cost $362
Lot 3:   300 units, unit cost $345
Lot 4:   400 units, unit cost $329

What are the learning curve variables (A, LC % and b)?

First let me show you how to set up this problem:

Recall the classic cumulative average learning curve equation:

$Y = Ax^b$

Those we have:

$Y_2 = \$362 = A(x+200)^b$

$Y_3 = \$345 = A(x+500)^b$ (the units for lots 2 and 3)

$Y_4 = \$329 = A(x+900)^b$ (the units for lots 2, 3 and 4)

Now solve these equations for A:

$A = \$362/(x+200)^b$

$A = \$345/(x+500)^b$

$A = \$329/(x+900)^b$

Now I am not going to use logarithms to solve this but rather I will use iteration on a computer. Why? Let's look at the first equation in log form:

$\text{Log } A = \text{Log}(362)/b*(\log(x+200))$

I have to take the log of x + 200 and that is not possible.

I am sure there is a mathematical way of solving this problem but I do not know what it is. I do know that given 3 equations and 3 unknowns I can find a solution.

What I did was set up a spreadsheet where I solved these 3 equations for a variety of units and learning curves.
Then I subtracted one set of calculations from another. My solution lays at the point where the subtraction of x2-x3 and x3-x4 is 0.

Here's the spreadsheet:

b:		Total	-0.02915
X1	X2	Units	98.00%
100	200	300	$427
200	200	400	$431
300	200	500	$434
400	200	600	$436
500	200	700	$438
600	200	800	$440
700	200	900	$441
800	200	1,000	$443
900	200	1,100	$444
1,000	200	1,200	$445

-0.05889	-0.08927
96.00%	94.00%
$507	$602
$515	$618
$522	$630
$528	$641
$532	$650
$537	$657

$540	$664
$544	$671
$547	$676
$550	$682

-0.12029	-0.152	-0.18442	-0.21759
92.00%	90.00%	88.00%	86.00%
$719	$861	$1,036	$1,252
$744	$900	$1,093	$1,333
$765	$931	$1,139	$1,400
$781	$957	$1,178	$1,456
$796	$980	$1,212	$1,506
$809	$1,000	$1,242	$1,550
$821	$1,018	$1,269	$1,590
$831	$1,034	$1,294	$1,627
$841	$1,050	$1,317	$1,661
$849	$1,064	$1,338	$1,693

b:		Total	-0.02915
X1	X3	Units	98.00%
100	500	600	$416
200	500	700	$418
300	500	800	$419
400	500	900	$421
500	500	1,000	$422
600	500	1,100	$423
700	500	1,200	$424
800	500	1,300	$425
900	500	1,400	$426
1,000	500	1,500	$427

-0.05889	-0.08927
96.00%	94.00%
$503	$611
$507	$619
$511	$627
$515	$633
$518	$639
$521	$645
$524	$650
$526	$654
$529	$659
$531	$663

-0.12029	-0.152	-0.18442	-0.21759
92.00%	90.00%	88.00%	86.00%
$745	$912	$1,122	$1,388
$759	$934	$1,155	$1,435
$771	$953	$1,184	$1,477
$782	$970	$1,210	$1,516
$792	$986	$1,233	$1,551
$801	$1,000	$1,255	$1,583
$810	$1,014	$1,276	$1,614
$817	$1,026	$1,295	$1,642
$825	$1,038	$1,312	$1,669
$832	$1,049	$1,329	$1,694

b:

X1	X4	Total Units	-0.02915 98.00%
100	900	1,000	$402
200	900	1,100	$403
300	900	1,200	$405
400	900	1,300	$405
500	900	1,400	$406
600	900	1,500	$407
700	900	1,600	$408
800	900	1,700	$409
900	900	1,800	$409
1,000	900	1,900	$410

-0.05889	-0.08927
96.00%	94.00%
$494	$610
$497	$615
$500	$620
$502	$624
$504	$628
$506	$632
$508	$636
$510	$639
$512	$642
$513	$645

-0.12029	-0.152	-0.18442	-0.21759
92.00%	90.00%	88.00%	86.00%
$755	$940	$1,176	$1,479
$764	$954	$1,197	$1,510

$772	$967	$1,216	$1,539
$779	$978	$1,234	$1,566
$786	$990	$1,251	$1,591
$793	$1,000	$1,267	$1,615
$799	$1,010	$1,283	$1,638
$805	$1,019	$1,297	$1,660
$811	$1,028	$1,311	$1,681
$816	$1,037	$1,324	$1,701

			X3-X2
X1	X2	X3	98.00%
100	200	300	-$11.76
200	200	300	-$13.49
300	200	300	-$14.67
400	200	300	-$15.54
500	200	300	-$16.21
600	200	300	-$16.75
700	200	300	-$17.19
800	200	300	-$17.55
900	200	300	-$17.86
1,000	200	300	-$18.13

96.00%	94.00%
-$3.67	$8.35
-$7.74	$1.15
-$10.55	-$3.86
-$12.63	-$7.58
-$14.23	-$10.48
-$15.52	-$12.81
-$16.58	-$14.73
-$17.47	-$16.35
-$18.22	-$17.74
-$18.88	-$18.94

92%	90%	88%	86%
$25.82	$50.77	$86.02	$135.48
$14.44	$33.89	$61.92	$101.92
$6.48	$21.99	$44.80	$77.90
$0.53	$13.04	$31.84	$59.62
-$4.12	$6.01	$21.61	$45.10
-$7.88	$0.31	$13.27	$33.20

-$11.00	-$4.44	$6.31	$23.23
-$13.63	-$8.47	$0.37	$14.69
-$15.89	-$11.94	-$4.76	$7.28
-$17.87	-$14.98	-$9.27	$0.77

			X4-X3
X1	X3	X4	98.00%
100	300	400	-$13.33
200	300	400	-$14.09
300	300	400	-$14.69
400	300	400	-$15.19
500	300	400	-$15.60
600	300	400	-$15.96
700	300	400	-$16.27
800	300	400	-$16.54
900	300	400	-$16.77
1,000	300	400	-$16.99

96.00%	94.00%	92.00%
-$8.67	-$1.15	$10.47
-$10.48	-$4.40	$5.24
-$11.93	-$7.04	$0.99
-$13.13	-$9.22	-$2.54
-$14.14	-$11.06	-$5.53
-$15.01	-$12.64	-$8.11
-$15.76	-$14.01	-$10.36
-$16.41	-$15.22	-$12.34
-$17.00	-$16.30	-$14.11
-$17.52	-$17.26	-$15.69

92.00%	90.00%	88.00%	86.00%
$10.47	$27.93	$53.69	$91.23
$5.24	$20.02	$42.18	$74.88
$0.99	$13.58	$32.75	$61.43
-$2.54	$8.19	$24.84	$50.11
-$5.53	$3.61	$18.10	$40.42
-$8.11	-$0.34	$12.25	$31.99
-$10.36	-$3.80	$7.12	$24.57
-$12.34	-$6.86	$2.58	$17.97
-$14.11	-$9.59	-$1.49	$12.05
-$15.69	-$12.05	-$5.16	$6.70

Now I do not have a single single 0 in either of my spreadsheets. But I do notice that at 90% in the X3-X2 (600+200+300) column I have $0.31. I also have -$0.34 in the X4-X3 (600+300+400) column. Those two values are the closest I have to 0 that match. Therefore I conclude that there were 600 units ahead of the three lots I now have and that the cumulative average learning curve is 90%.

You cannot use this same trick with only two lots because there will be multiple learning curves and As that will solve that problem.

Since I used a 90% curve and 600 units to gather the data for the problem I am satisfied that this method works fairly well. As always with learning curves, you should use it with caution since it is a brute force method and not a closed form solution. Still it can be done and given the right set of circumstances you can pick up a side bet or two on this.

If you wish to determine the unit learning you can either convert or else work the problem using midpoints instead of the last unit in the lot.

Chapter 30 Will the real A please stand up

One of the most vexing problems with using lot data to determine the learning curve parameters is the fact that the A value won't behave. Let me show you an example:

I have 3 lots
Lot 1: 400 units, unit cost $850
Lot 2: 500 units, unit cost $750
Lot 3: 400 units, unit cost $630

What is the unit learning curve?

Well let's solve the first two lots and then look at the change when we solve for all 3 lots using the average cost method.

 The first two lots yield the following learning curve parameters:

A: $1,284
LC: 94.39%

Now when I solve for all 3 lots I get the following:

A: $1,485
LC: 92.45%

Now it makes sense to most people that the learning curve will change but shouldn't the A value stay the same?

Well let's see what would happen if I used my A value and learning curve from the first two lots to predict the three lots and then I will compare it to the three lot prediction parameters to see which is best. I will again determine the absolute deviation and define the one with the least absolute deviation as the best fit.

Learning Curve parameters based on the first two lots

LC:	94.39%
b:	-0.08329
A:	$1,284

Units	Actual Cost	Midpt	Predicted Cost
400	$850.00	142.56	$849.47
500	$700.00	632.54	$750.33
400	$680.00	1,093.89	$716.86

Absolute Deviation	Total Deviation	Total Cost
$0.53	$210.64	$340,000.00
$50.33	$25,163.87	$350,000.00
$36.86	$14,745.44	$272,000.00
	$40,119.95	$962,000.00
Error	4.17%	

Now what happens when I predict all three lots using the three lot prediction parameters?

Learning Curve parameters based on all three lots

LC:	92.45%
b:	-0.11325
A:	$1,485

Units	Actual Cost	Midpt	Predicted Cost
400	$850.00	140.32	$848.31
500	$700.00	632.03	$715.37
400	$680.00	1,093.71	$672.29

Absolute Deviation	Total Deviation	Total Cost
$1.69	$675.75	$340,000.00
$15.37	$7,684.02	$350,000.00
$7.71	$3,084.24	$272,000.00
	$11,444.01	$962,000.00
Error	1.19%	

The error rate has fallen significantly. Remember that we are using lot data as a proxy and since we are, in effect, developing a best fit curve for our data based on both the A values and the b values then it would be surprising if the A value did not change.

Which is the real A value? Well there isn't one. If there was a real A value that the contractor was following then there would not be a change in the A value when we use lot data to determine our values. For instance, assume the following data set:

Units
400 $567.84
500 $450.33
400 $414.26

Now determine the learning curve parameters for a 2 and 3 lot case.

The two lot case gives the following parameters:

A: $1,189
LC: 90.09%

The three lot case gives the following parameters:

A: $1,190
LC: 90.08%

I can readily see that this contractor is coming down a 90% learning curve and that the A value is going to be about $1,200.

When it is important that a learning curve be accurately predicted (for instance when you have committed to a certain learning curve for pricing or for your evaluation) you should be sure that your firm and the other firm understand which A is going to be used in the calculations. I prefer to take the A value from the first two lots and use that to determine my progress but that is just a personal choice. As long as the contractor is not actually using a learning curve for pricing (and he isn't in the above example or else I would not get such a wide variance in learning curve parameters) then it is impossible for the real A to be determined.

Chapter 31 When is learning not learning?

Here's a situation:

First lot: $8,00,000 Units: 1,200
Second lot $8,000,00 Units: 1,500

What are the learning parameters?

You can easily determine the unit learning curve from methods I have illustrated elsewhere in this text:

The short answer is:

LC: 90.42%
A: $15,958

QED.

Then something comes along and bursts your bubble. The first lot included $2,000,000 for tooling and non recurring engineering, the second lot contained $1,500,000 for tooling and non recurring engineering.

Usually tooling is bought early in a program. It is not uncommon that all the tooling required for a program will be purchased in the first 2-3 lots. It is true that there will be maintenance requirements for the tooling and that there may be better tooling that comes along that will be purchased but these are not learning curve issues. To determine the true learning curve (in my not so humble opinion) tooling should be excluded from the learning curve analysis.

Non-recurring engineering is a catchall term used in defense cycles. It covers a sundry of things such as trade studies, methods analysis, special studies, building of models etc. There is little if any learning that goes on in these types of activities. In fact, NRE will fall greatly after the first couple of lots. Again in my not so humble opinion these costs should be excluded from the learning curve analysis.

Now let's see what happens to our unit learning curve after we account for these activities:

 First lot: $6,00,000 Units: 1,200
Second lot $6,5000,00 Units: 1,500

My parameters become:

LC: 93.63%

A: $8,869

If I am being held to a specific unit learning curve, I will want to keep the tooling and NRE in my calculations because it sets a higher A value and it will allow me to more easily reach my goals. Let's assume I am going to purchase a total of 15,000 units and I can use either method to determine whether I reached my goal of a 90% learning curve.

$$Y_{avg} = A(x_{mid}^b)$$

For the first case:

$$Y_{avg} = \$15,598 \, (5,074.42^{-0.152}) = \$4,264/unit$$

For the second case

$$Y_{avg} = \$8,869(5,074.42^{-0.152}) = \$2,425 /unit$$

Usually a contract will define what is and is not part of the learning curve for various purposes. If a contract is using only total contract costs as a guide then it is assumed that all costs (tooling and NRE inclusive) will be used in the determination of the learning curve for evaluation. If I am a producer, I am very happy about this. As a consumer I would have my doubts.

Now it is true that if someone was foresighted and took the time to develop a learning curve goal based on all costs (tooling, NRE and everything including the kitchen sink) and from that developed a learning curve goal, then the math will all work out in the end. We would have a different goal when tooling and NRE are included, that's all. But usually the language will be much less specific and just say that the contractor will try to achieve a 90% (80% or whatever) unit (or cumulative average) learning curve. So check your contracts closely to determine what it says. The terminology may or may not work to your advantage. If it doesn't, you should try to work an example to show the managers why the terminology is deficient.

Chapter 32 Cost trade offs in methods selection

When you are developing processes and methods, it often becomes evident that (when there is substantial production) the more highly productionized (or mechanized) a process or method, the less expensive the unit cost. The question for management becomes when should I make the investment in the production equipment.
Let's set up an example:

I am building widgets
I have a 90% unit learning curve operating
My A value is 4 hours
I can purchase a machine that will allow me to build widgets for ½ hour each at a cost of $150,000
The machine will have a life of 50,000 widgets and then must be replaced
I am going to ignore the tax consequences (although you shouldn't and if you wish to explore this further you should obtain a book on engineering economy)
My labor cost is $65 per hour

Maintenance cost on the machine will be $20,000 over the 50,000 widgets

I need to know how many units that I must build in order to switch from the manual method of building the widget to the machine driven method.

The unit capital cost for the machine driven process is: $150,000/x

The labor cost is ½ hr * $65/hr = $32.50 per unit

The unit maintenance cost for the machine driven process is $20,000/x

The per unit cost for the machine driven process is:

$Y = \$150,000/x + \$32.50/unit + \$20,000/x$

The unit cost for the manual process is: Y = $A(x^b)$ * $65/hr
Where A = 4 and b = -0.152

Set these equal to each other

$\$150,000/x + \$32.50/unit + \$20,000/x = 4(x^{-0.152}) * \$65$

Now solve for x:

Since we have a linear equation on the left hand side and a non-linear equation on the right hand side, I do not know how to take logarithms to solve so instead I used a spreadsheet.

Solution for cost tradeoffs example:

x	4,175 Number of units
Per unit	
$150,000	$35.93
$32.50	$32.50
$20,000	$5
	$73.22 Machine method cost per unit
LC:	90.00%
b:	-0.152
	$65.00 Cost per hour
A	4
	$73.22 Labor method cost per unit
	$0.00 Delta

At approximately 4,175 units the cost to purchase and maintain the machine is justified by the extra labor cost. Here's the worked equation:

$$\$150{,}000/4{,}175 + \$32.50/unit + \$20{,}000/4{,}175 = 4(4{,}175^{-0.152}) * \$65$$

Solve for either side and you should get something very close to $73.22 per unit.

This calculation will allow management to make a decision as to when (if ever) to productionize the process. For some buys, it makes no sense to purchase very expensive equipment because there simply are not enough units being purchased to allow the fixed cost versus variable cost trade off.

This analysis can also be used to show how much more effective a machine process is than a manual process. For instance, let's say that we are going to produce 30,000 units. What are the costs for both methods and what are the savings for using the machine process?

Solution for cost tradeoffs example 2:

x	30,000	Number of units
	Per unit	
$150,000	$5.00	
$32.50	$32.50	
$20,000	$1	
	$38.17	Machine method cost per unit

LC:	90.00%	
b:	-0.152	
	$65.00	Cost per hour

A   4
$54.25 Labor
method
cost per
unit

$16.09 Delta
$482,640 Total
Savings

The savings at 30,000 units is appreciable (almost $1/2 million).

This analysis has focused on a few simple assumptions and there will be many more when an actual decision is made (tax implications, inflation, probability of obtaining more business, etc). The point of the example is to demonstrate that there is a process whereby you can develop a mathematical model to determine what method would work best for a given set of circumstances. The cost engineer and the industrial and production engineer should work hand in hand when developing proposals for upper management decision.

Chapter 33 What Negotiators and Managers Should Know about learning curves (and usually don't)

This section will provide some practical advice to managers and (particularly) negotiators. As I have acidly pointed out many times in this text, personnel that do not understand learning curves often negotiate bad situations for a firm because of this lack of knowledge. Don't worry about the math, understand the concepts.

First, make sure you know what type of learning curve is going to be utilized either in the contract or in the cost tracking. If one side is doing budgeting and forecasting based on cumulative average curve and the other side is doing the same activities based on a unit curve there will be serious differences in opinions as to future costs.

If you are a producer make sure you know what learning curve(s) are going to be used by the other side to evaluate your performance. If the customer is going to use a 90% unit learning curve for all costs and you are using 95% on some and 85% on others there will again be differences in future cost projections. Keeping these differences to a minimum makes for a happy customer and a happy producer.

Try to determine how the customer/contractor is going to determine the A values. Remember that the A value is normally a mathematical construct and it is normally estimated and then later taken from lot data once more than one lot has been negotiated.

Try to ensure that both sides knows what will and will not be evaluated according to learning curves. Often tooling and non recurring engineering costs are excluded from learning curve analysis, though sometimes it is not. Every dollar that goes into the analysis should be acknowledged as either a learning curve item or a nonlearning curve item.

From a practical perspective, it is important to understand what evaluation techniques are going to be used to determine if the contract is successful. If the evaluation is based on a total cost value then the learning curve percentage may not matter. Often though the learning curve percentage is one of many factors that are used to determine the success of the contract and this information should be known by the contractor.

It is extremely important that you understand how inflation will be handled. I think this is one of the common misunderstandings in the use of learning curves by managers and negotiators. A 90% learning curve is based on constant dollars and/or labor hours. Failure to consider inflation when calculating the proper learning curve will underestimate the learning and will cause problems on both sides of the equation.

Determine how the contractor will be paid or how you will be paid if you are a purchaser. Are there incentive clauses that will be paid if you meet certain criteria (such as a certain cost or a certain learning curve)?

To make things clear, I suggest that both sides provide mock spreadsheets that show how they would forecast costs. If both sides have their technical experts review these forecasts, an agreement can be reached on how best to track costs and everyone benefits.

My best advice is to make sure an experienced cost engineer/analyst supports the negotiation effort. Often, negotiators will make concessions on both sides without realizing the practical impact on the organization. For instance, committing to an 85% learning curve versus an 88% learning curve does not sound to significant, but it is very significant when trying to obtain those goals.

Chapter 34 Things that cause learning problems

There are more things that cause learning problems than I care to list but here are a few of the more important ones.

1. Poor Quality of parts. If the parts cannot be made to specification then the learning curve cannot operate properly. There will be many parts that contribute to the learning curve that will be scrapped. The learning is still ongoing but the record keeping will not reflect this learning.

2. Poor methods. This is without doubt the largest contributor to learning curve problems. A poor method will often require an operator to do things with the wrong hand tool and at the wrong time. The best investment a firm can make is to hire a good engineer to not only develop the method but a different good engineer to check the method with some trained operators prior to putting the method into operation.

3. Computer time. In this day of computers, the work cell is awash in computers. Computers can make the job easier and quicker but they can also slow the time required to do a task appreciably. Why? I'll give you an example:

I was assigned the task of time studying a work cell that was having significant problems in meeting the projected rate. The first thing I discovered was that over half their time was spent in updating data into a computer system. The original rate projection had not even considered computer time in their calculations. What had happened was that the methods engineer and the time engineer did not talk and as a result the company was now paying the price. During our investigation of the computer time we found that most of the stop points were inserted as a quality assurance measure, the operator was supposed to visually inspect the product and then update the computer that he/she had done so. Just by taking out these unnecessary steps we cut the computer time by 75%.

4. Employee overloading. You will often find work cell employees answering phones, fetching supplies, opening

secure doors, sweeping the floor and a 1,001 other small things. These things add up to real time. The employees should be protected from all ancillary but necessary jobs and instead they should focus on performing the productive labor they were hired for.

5. Failure to pass around information. Often one employee will come up with a superior method of doing a job but no other employee will know about that method. A firm should have a suggestion program or some other method (quality circles for instance) of having operators pass on information about production and quality issues. The operator knows the process and knows more short cuts than anyone else in the firm. Use this resource and it will pay huge dividends.

6. Management indifference. Sounds impossible right, upper management must care what the production is because that's how they are measure. Not always. I have found that for certain types of contracts, having a good learning curve actually costs the firm money. Maybe management indifference is the wrong title but rather contract indifference. If it

makes more financial sense to have a poor learning curve (as it does in some cost plus contracts) than to have a good learning curve then you can bet the firm will achieve a poor learning curve and will have some great excuses as to why they could not achieve projections.

7. Design changes. Design changes can be a killer as far as a consistent learning curve is concerned. Just as the operators are getting the hang of something, along comes a substantial change and back up the learning curve you go.

Chapter 35 Previous learning and Complexity Factors

A common cost estimating technique is to take a known item cost (or hours) and to use a complexity factor to estimate the cost for a new application. Let's say for instance, that the cost for a widget used in Part A is $500 at unit 8. I am building part B which will also use a widget, however this widget is slightly larger and will therefore require a redesign. How do I estimate the cost of the Part B widget? Well here's one method.

Ask the design engineers just how complex the new widget will be. Let's say they estimate it will be at least 25% more difficult to build than the current widget. Now I have a known cost and a complexity factor and I can arrive at a defendable (although not necessarily very accurate) estimate. Here are the computations.

Widget cost = $500 at unit 8

$$Y_8 = Ax^b = \$500$$

I know from my past experience that I am operating to a 90% unit learning curve there fore $b = \log(.90)/\log(2) = -0.152$

The cost for A becomes:

Log ($500) = Log A $-(0.152)*$ Log (8)

$2.6989 = $ Log A $- 0.1373$

2.6989+0.1372 = Log A

2.8361 = Log A,

A = Antilog (2.8361) = $10^{2.8631}$ = \$685.64

Let's check this to see if it worked:

$Y_8 = Ax^b$ = \$500,

Y_8 = \$685.65 $(8^{-0.152})$ = = \$500

So it checks:

Now apply the complexity factor:

A for widget for new part = 1.25 * A for widget for old part:

\$685.65 * 1.25 = \$857.06

Now you have an A value and an assumed learning curve (90%) and you can therefore make any projection you would like to make for the new part.

A common question that arises during the use of complexity factors is: Should I apply prior learning to the new part that I am developing. For instance,

If I have produced 8 parts of widget A, should I start the learning curve for widget B at part 9?

Like most questions of this sort the learning is in the eye of the beholder and one can probably argue either way. For instance, if there are similarities then probably some learning will attach to the new part but how does one account for it without real data. I try to avoid the problem entirely by starting over the learning for the new item. Engineering estimates are more often low than high and I like to be able to say that I have conservatively estimated the work.

Chapter 36 Wrapping it up

If you take nothing else from this book remember these things:

Learning curves are forecasts they are not written in stone. The formulas used to calculate the learning curves are always wrong but they are better than anything else we have yet developed for predictions.

Learning curves are great for helping the decision makers determine what problems are occurring and where they are occurring.

Learning curves should be used with caution when a firm is committing to a specific learning curve. It is very important that both sides understand how the all parameters (particularly the A value) will be calculated and how inflation will be handled in the learning curve.

When all else fails, take the raw data and brute force the classic learning curve equations on a computer. Even if you do not find the "right" answer, you will still obtain some insight into the process.

It is better to turn on the lights than to curse the darkness, so gather **DATA.** You can't tell the players without a score card and you surely can't determine what learning curve is operating without having the right data in the right form to make that determination.